# A
# CROWNING
## SYNDROME
that changed the world order

**MIKE RANA**

INDIA · SINGAPORE · MALAYSIA

**Notion Press**

Old No. 38, New No. 6
McNichols Road, Chetpet
Chennai - 600 031

First Published by Notion Press 2020
Copyright © Mike Rana 2020
All Rights Reserved.

ISBN  978-1-64899-614-6

# A Crowning Syndrome

that changed the word order

**The virus speaks**

Those who perished will not return
Others are waitin' to die
I'm just a microscopic virus
Deadly that no man can defy

My crown became your nadir
Till death do us part
Seasons may come and seasons may go
But I, go on forever

If you dare to disobey, you can
If you dare to play, you can
Nature will have its say
But handle me now, if you can

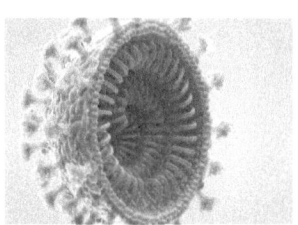

## Disclaimer

This book is more like chronicles than a book because things about symptoms, vaccines, and the number of deaths are changing on a daily basis. Things were only settling down when the book was taken up for writing. Much of the data or facts might have changed by the time you get this book in your hands.

The manuscript was submitted for printing on **20-May-2020**

Substantial quotes have been made from different and well-respected sources. Nothing in this book is written with prejudice in mind. The references are for fair use, and to share knowledge. Should anything be noted as a mistake please accept my apologies.

It is necessary to give a caution; many words used in this book are used in their simplest of sense that a common man can understand.

# 1. **Foreword**

Corona is here to stay. It touches all aspects of our life and all humans around the world. It disrupts day-to-day affairs, friendships, relationships, work culture, education, and health, including all the rest that happens in your mind. The optimist may still remain cool and learn about new ventures but the pessimist will empty one's internal worst.

This novel virus places unknown anxiety on the society as a group, and untimely negative stress on the economies of countries by impacting international trades, commerce, and even mutual relations. The wicked pandemic has affected more than half of the world's population. In terms of expanse, impact, and overall negative consequences, it may prove to be more lethal than the two World Wars in the last century.

It's now a substitute for the third World War short of bullets fired. It pushed everyone inside homes instead of taking shelters in underground bunkers urging taxpayers and their masters to look back at the money spent on weapons, aircraft, and warships instead of healthcare systems. **It makes one wonder what priorities the governments have set for us.** Religious wars and political scuffle seem like vile catfights when people face deadly virus amongst their personal lives.

They say the virus will remain latent forever and will keep reminding us that something is hidden inside our bodies that need to be expelled. You are lucky if you can maintain your cool.

The world's best minds are at it to understand what the hell it is. We smartly race with each other to find a cure to conquer it. But there are lingering doubts in our minds. Are we driven to our doom? Is this earth's way to take a breather from relentless exploitation by human beings?

Mike Rana has worked hard to present authentic information before the reader gets a clear idea about the challenges ahead.

Puneet Mehra        Gurugram (Haryana)

Mohini Sharma        Kullu (HP)

Cover design

Tanushree Rana

## 2. Preface - Adieu People

Let's pause a minute, pay tributes, and shed a tear for all those who perished in this unfathomable disaster.

### Epitaph

*Gone but never forgotten*

*Rest in peace, you will be missed*

*You were a friend to many, and stay as you were*

*Your life will be measured in memories*

*The summer's come and you are gone*

*Nothing will ever take your place*

*Until we meet again, and in this world*

Consider our life in the backdrop of Corona – the virus supposedly crowns of light now turning slowly into a nadir of darkness. Are our bodies so vulnerable that all our best minds and our views about science, religion, genetics, and economy are simply toppled when something critical like this happens?

# 3. **Acknowledgements**

Quite a lot of quotations, data, texts, and YouTube links have been used in this book. These were required for authenticity since everything about COVID-19 is full of qualms and misgivings. Footnotes have been inserted at appropriate places to make easy reading.

I place on record my thanks for the knowledge of various people and for using their ideas for a 'Fair Use'.

1. Wikipedia
2. Google chrome
3. Centres for Disease Control and Prevention (Federal agency) headquartered in Atlanta, Georgia.
4. The World Health Organization (WHO/OMS) headquarters in Geneva.
5. The Ministry of Health and Family Welfare, India Founded in 1976 HQ New Delhi, India
6. Life Science Journal Jeanna Brynner Editor-in-chief, June 14, 2018
7. Nature Research - a division of the international scientific publishing company Springer Nature, that publishes academic journals, magazines, online databases, and services in science and medicine. Nature Research's flagship publication is Nature, a weekly multidisciplinary journal first published in 1869. HQ Berlin, Germany
8. OWID - Our World in Data is a scientific online publication that focuses on large global problems such as poverty, disease, hunger, climate change, war, existential risks, and inequality. Led by Max Roser
9. The University of Ottawa in Canada

# Contents

## 4. As the COVID-19 [1] takes charge

News about [2] coronavirus is viral, and it doesn't stop even after months of being viral, rather it became more aggressively viral. The media is overwhelmed with deaths or those who are in the realm of death. We can only mourn them and for the rest, the number of survivors is very encouraging.

It's a novel virus, meaning no one has built up an immunity to it. Its spread is exponential. And its cure beyond comprehension. Period.

*Anthony Stephen Fauci* [3] is seen as a trustworthy voice, in the eyes of the US general public, and who stood by the side of US President George Bush in 2005, when Bush said, 'America doesn't have a pandemic now but trust me when it starts, the pandemic doesn't recede. It survives to kill until no one is left to count the dead'. Historically both of them were right.

### The outbreak occurs

The newsbreak was sudden. We woke up a wicked morning with the news that people from all over the globe had been affected and taken ill. Some dangerously, others seriously, and some died. Some unfortunate ones were knocking at the doors of death. All infected by a so-called virus, whose symptoms initially looked similar to influenza. But that was an easier comparison which lasted only a few days.

---

[1] CO for corona, VI for virus, D for disease and **19**

[2] 'Corona injects a dose of illiteracy to literates and confuses both', **Mike Rana**

[3] **Anthony Stephen Fauci**, is an American physician and immunologist who has served with a large number of US presidents, (as the director of the National Institute of Allergy and Infectious Diseases since 1984)

It was a virus call that made us all glued to the television. It forced us to read what we would've never read hitherto; magazines, journals, websites (official and cursory), and I believe it was not the end of the story. It was just a start.

The first known case of COVID-19 was a 55-year-old from Hubei province in China, as reported by the South Morning China Post and reported by Live Science journal. That case dates back to 17-Nov-2019.

According to al Jazeera, the week 24-Feb-2020 to 01-Mar-2020 marked the confirmation of first cases in countries across the world, including Kuwait, Bahrain, Iraq, Oman, Qatar, Norway, Romania, Greece, Georgia, Pakistan, Afghanistan, North Macedonia, Brazil, Estonia, Denmark, Northern Ireland, and the Netherlands, Lithuania, and Wales.

On 24-Mar-2020, a lockdown was declared in India. I, like the many others, came under its aftermath in my home. I believed it was temporary and the only plausible option if we were to escape the wrath of the virus.

I am lucky that I don't go to any office, my desktop at home suffices all my needs including the writing of this advisory kind of book. And it works if the internet works. So, in effect, nothing practically changed in my case.

Symptoms of COVID-19

Perhaps my most arduous task was to logically and comprehensively list down all the symptoms that one could experience in this sort of malady.

Most COVID-19 infected people experience mild to moderate respiratory illness and recover without requiring special treatment. Older people and those with underlying medical problems like cardiovascular disease, diabetes, chronic respiratory disease, and cancer are more likely to develop serious walkovers. And their return to normal is dubious.

The virus didn't change gears, it was simply going by the location of *angiotensin-converting enzyme 2* (ACE2) receptor in human bodies. If ACE2 was stronger in an organ the Covid-19 manifested in that organ for that patient. The virus didn't become more lethal as it is being reported these days, it is the result of more data being made available as more and more patients appeared.

Simple measures - such as washing your hands, disinfecting the frequently touched surfaces and objects, and avoiding touching your face, eyes, and mouth - can greatly lower your risk of infection.

A small official comparison of symptoms may help.

**Reference table from CDC**

| Seasonal Flu | COVID-19 |
| --- | --- |
| Happens annually and usually peaks between December and February | Rarely happens (three times in 20th century) |
| Usually some immunity from previous exposures and influenza vaccination | Most people have little or no immunity because they have no previous exposure to the virus or similar viruses |
| Certain people are at high-risk for serious complications (infants, elderly, pregnant women, extreme obesity and persons with certain chronic medical conditions) | Healthy people also may be at high risk for serious complications |
| Vaccine available for annual flu season | A vaccine may not be available in the early stages of a pandemic (12 to 18 months required) |
| Usually, one dose of vaccine is needed for most people | Two doses of vaccine may be needed |

**Comparisons by the others**

| Symptom | Malady |
|---|---|
| dry cough + sneeze | Air Pollution |
| cough + mucus + sneeze + running nose | Common Cold |
| cough + mucus + sneeze + running nose + Body ache + weakness + light fever | Flu |
| Dry cough + sneeze + Body ache + weakness + high fever + difficult breathing | corona virus |

**Silent Hypoxia**

An interesting and rather scary description appeared on a personal blog of someone. It stated, 'When I started feeling that I was in it, a kind of electric current gushed through my body touching all my organs as if my blood was circulating rather awkwardly along with some strange constrictions. And it terminated in my throat with a choking sensation like I was suddenly placed at very altitude and oxygen was scanty. I was getting choked, my body was turning blue'.

These are the signs of hypoxia that a flier may feel when he makes a loop on his jet at very high altitudes and supersonic speeds.

On the ground level, patients whose blood oxygen saturation levels are exceedingly low but who are hardly gasping for breath could get these symptoms. Normal blood-oxygen levels are around 97%, Dr Marc Moss said, and it becomes worrisome when the numbers drop below 90% and the brain may not get sufficient oxygen. Patients might start experiencing confusion, lethargy, or other mental disruptions. As levels drop into the low 80s or even below, the danger of damage to vital organs rises.

*Silent hypoxia* may be killing COVID-19 patients, but the virus is not the main cause of their troubles says, Stephane Pappas. [4]

**Mysterious blood clots in COVID-19 patients**

Many doctors have recently reported seeing an alarming number of COVID-19 patients with blood clots, gel-like clumps in the blood, that can cause heart attack and stroke, according to news reports.

'The number of clotting problems I'm seeing in the Intensive Care Unit (ICU), all related to COVID-19, is unprecedented,' says Dr Jeffrey Laurence. [5] Some doctors started to notice that their COVID-19 patients were developing clots in their legs, even while they were on blood thinners, according to The Washington Post.

**Chicken Pox look alike [6]**

This is the latest and possibly weirdest COVID-19 symptom according to the Johns Hopkins University.

Researchers in Spain have found purple-coloured lesions from five to 15 millimetres in diameter (similar to chickenpox) on the feet on COVID-19 patients including children, adolescents, and adults.

The first documented case of the lesions was found on a 13-year-old boy in Spain who had the foot lesions before he exhibited any other symptoms. Doctors have since found other cases that indicate that foot lesions can be the first symptom of COVID-19, appearing before others such as a sore throat and cough.

---

[4] **Stephanie Pappas** - Live Science Contributor

**Dr. Marc Moss**, the division head of Pulmonary Sciences and Critical Care Medicine at the University of Colorado Anschutz Medical Campus.

[5] **Dr. Jeffrey Laurence**, a haematologist at Weill Cornell Medicine in New York City

[6] statement from the **Spanish General Council** of Official Podiatrist Colleges.

## Brain Dysfunctions

Other than the usually known symptoms of COVID-19 brain diseases or dysfunctions were seen.

## New York peculiarities

Nearly 1 in 7 people in New York who were randomly tested for coronavirus antibodies turned out to have them, Governor Andrew Cuomo announced on 23-Apr. In New York City, the number is even higher, about 1 in 5 people tested positive for antibodies to SARS-CoV-2.

If those early results translate to the rest of the New York population, that would mean about 2.7 million people across the state would have been infected. Shocking. No doubt the New Yorkers are on the peak of this pandemic.

The antibodies suggest these people were exposed at one point or another to the coronavirus and recovered, Cuomo said. However, it is still not known whether or not these people are now immune to it.

## Disease without visible symptoms

This virus is deadlier, more infectious, and catastrophically, it sends far more and more quickly, ill people to the hospital, than all other maladies put together. Regrettably, the confirmation of infection comes after the stated 2-14 days after exposure. For people with less immunity the accumulated symptoms may show signs even earlier, it is so stated.

And the variable elapsed time, from 1 to infinity, for its detection, is driving us nuts. One doctor stated, 'it might affect all the people one time or another in their life, just like flu'.

The most worrisome case is when a person is not COVID-19 positive and has been latently exposed. He further goes on to infect the others including his family members, near friends and co-passengers, and those who come in his contact even in another city or country.

All of them the *primary*, *secondary* and *tertiary* members of this expanding or growing tree, become suspects of the Corona, and each one could adopt the role of the primary for further expansion. All without his knowledge or the other persons of the collection.

The only indication comes when he or some third person notices him coughing, sneezing, or breathing uncomfortably.

But this is no way to identify any serious illness. We need suitable testing mechanisms.

**Viruses versus Humans**

Will viruses displace we the human beings?

One of the biggest factors allowing COVID-19 to continuously infiltrate populations is its ability to spread silently via asymptomatic carriers. Most people who come into contact with COVID-19 won't end up developing symptoms. However, those asymptomatic *carriers* are still quite contagious.

One would assume that COVID-19's inability to bring about symptoms in many carriers represents a deficiency or weakness in the virus. A team of researchers at Princeton University *disagrees*.

According to them, the asymptomatic transmission is a logical evolutionary step for pathogens. It's become harder and harder for pathogens and viruses to thrive like they did hundreds of years ago. It makes sense that new viral strains would eventually develop the ability to spread secretly via asymptomatic carriers.

It's a chilling concept, but an important realization: we've gotten smarter, and now pathogens have as well.

After a complex analysis of how pathogens can spread through populations, they found that asymptomatic transmission can be a successful *strategy* for viral survival.

4-15

We don't usually place viruses and pathogens on the same level as animals or ourselves, but just like more complex organisms, viruses are subject to natural selection. Viral traits that help strains survive and spread persist, while characteristics that don't aid survival die out eventually.

For example, if a virus were so powerful that it immediately killed its host before that organism had a chance to infect anyone else, that virus isn't going to last very long.

COVID-19 is quite mild, in many patients it has helped its spread all over the world. It's impossible to keep track of all asymptomatic COVID-19 carriers, meaning there are almost certainly countless people venturing outside today and spreading the coronavirus to many other people.

'Viral evolution involves a trade-off between increasing the rate of transmission and maintaining the host as a base of transmission.' Species that navigate this trade-off more effectively than others will come to displace those others in the population. In many ways, the relationship between any virus and its host is parasitic. The virus needs its host to stay alive, at least for a certain period.

Is the virus going to destroy humanity in favour of some others species, while evolution of species takes place?

**Threat at your doorsteps**

In the past, the phrase 'Hannibal is at the gates' was used to signify a sense of doom and imminent danger. Hannibal was a Carthaginian general who won many victories against the Romans in the Punic Wars. The 'Gates' referred to as those of the Roman Empire's capital. His armies were metaphorically considered right outside the city.

Likewise, COVID-19 is the looming threat at our doorstep, we step out we may invite her to come in. We can't call it the arrival of doom's day and if we continue to neglect, it might become.

## Actions and reactions on COVID-19

All is currently set for a Herculean effort on part of agencies, medical research, journals, and the world leaders, that they have a job to do and they better do it before we write our epitaphs. If we remember the 'Pearl Harbour moment' and the '9/11 moment,' Corona is not going to be localized. It's much more encompassing.

To enable your appreciation of how dangerous it could be to openly talk about COVID-19, the US Captain Brett Crozier was fired after a letter he had written to the Navy - outlining the impossibility of social distancing on the aircraft carrier - was leaked to The San Francisco Chronicle, according to the Times and a Live Science report. 'We are not at war. Sailors do not need to die. If we do not act now, we are failing to properly take care of our most trusted asset - our sailors,' Crozier wrote in the report. And I believe he was right and his castigation wrong.

As a writer is it a wrong question to ask, 'why are we in such a state of unpreparedness when such calamities did introduce themselves to us in our past?' Why don't we possess the required hospitals, the desired facilities, the test kits, and the medical staff to take care of this illness? And we are running helter-skelter in our efforts to contain this microscopic monster!

What you have read till now is all we know and then there is the unknown.

As the curtain unfolds, I recollect a Hollywood 2011 movie 'Contagion' directed by Steven Soderbergh, which I had seen a few months back on Netflix. The disease spreads to humans through contact with infected pigs in Malaysia and Singapore, and through consuming a fruit that was contaminated with urine or saliva from infected bats in Bangladesh and India. And I also remember Kate Winslet her heroine, died two days later just after Contagion, in her family home in suburban Minneapolis. She collapsed with seizures.

Her husband, Mitch, rushed her to the hospital, but she died of an unknown cause.

## The Contagion Movie

The movie reviews included a comment, 'Stephen King did the deadly virus-meets-the-supernatural thing with *The Stand*. Steven Soderbergh matches it up with something even scarier: cold, hard, medical reality'.

I wouldn't hope this movie manifests itself as truth in real life but the signs are that it has and it will. All eyes are hopefully set on the WHO and CDC to find a vaccine. It is a different matter altogether that even after elapse of many days, WHO/CDC combine failed to intervene or even raise an alarm as they had done during the EBOLA outbreak.

## What Bill Gates says – Historically [7]

Bill Gates is no ordinary man, particularly in the eyes of a computer professional. How can we forget his feat that sitting in his garage he

---

[7] **Bill Gates** read the January 1975 issue of popular electronics which demonstrated the Altair 8800, and he contacted Micro Instrumentation and Telemetry Systems (MITS) to inform them that he and others were working on a BASIC interpreter for the platform. In reality, Gates and Allen did not have an Altair and had not written code for it; they merely wanted to gauge MITS's interest. MITS president Ed Roberts agreed to meet them for a demonstration, and **over the course of a few weeks** they developed an Altair emulator that ran on a minicomputer, and then the BASIC interpreter.

Bill Gates was partnered with business magnate George Soros in owning a lab in Wuhan, China (the presumed location where the COVID-19 outbreak originated). And it is said that he is partnering most of the companies of the world selling vaccines. No doubt he has business interests and a lot to gain from COVID-19. Not a sin but immorality may be.

completed the writing of the Basic language interpreter and the base MSDOS that subsequently became IBM PC DOS operating system?

I want to draw your attention to his TED talk a few years ago. He stated the following. 'For donkey's years we, and the world, has been building bunkers for protection against a looming nuclear war and spent a hell of a lot of money. We built bunkers where we could shelter and save ourselves, eating, and drinking through cans. In Israel, there are at least four corner rooms on every floor in heavily populated buildings where one could hide when nuclear, chemical, gaseous warfare begins'.

I agree and most of you will, we should be ready with our medical defence systems; appropriate shelters against future viral attacks. With a rough guesstimate, next time, even this time, it could affect 20 Million or more humans around the world.

### How the others reacted to this suggestion [8]

*Dr Courtney Gidengil*, a senior physician policy researcher at RAND Corp. and a paediatric infectious disease specialist in Boston, said, 'in a perfect world if we could undertake these steps aggressively and immediately that would give us the best chance of crushing the curve. But the big question is how feasible it is to get these things to happen quickly enough to make a difference.'

Another major challenge is the risk of having the virus come back into a country from another part of the world. 'In terms of *truly* containing it, we need high levels of immunity against the virus,' Gidengil said. Immunity could come through a vaccine (which in all probability is at least a year away) or through recovered patients.

*David Hutton*, an associate professor of health management and policy at the University of Michigan School of Public Health, agreed that to really *defeat* this in the long term and get back to 'business as usual', we will need a highly effective treatment or a vaccine. 'Until

---

[8] Live science journal

the virus is controlled on a global scale *constant vigilance* is needed,' he said.

Hutton does think it's possible to re-open the economy by June, in line with what *Anthony Stephen Fauci* opines. As China is beginning to do. But he noted also that strict measures are in place in China, including temperature checkpoints and surveillance applications on people's phones. Such measures could be more challenging in the United States, where people are less willing to share private information, he said.

**Are we ready? Probably no.**

For disasters, we should work like Hollywood movies do, like what Bill Gates says. The film director puts everything in place, adequate surveillance systems, support systems, hundreds of workers, and above all the medical teams activated. Everything is just ready and on-call for the heroes.

The world has mobiles to communicate and we can use them to broadcast an early warning. We don't need a radar for that. We can pass instructions or seek reports, use satellite maps to see where people are and how they are moving around, and medical experiment in labs to develop the vaccines.

But this is not the way things work in real life. We may be able to get things organized, maybe in bits and bytes but an important question remains. What are our systems and processes, and are there any systems at all, and can we deploy them quickly? Can we bind them to work in unison? Otherwise, how we wish we had the luxury of a quick vaccine or one could take blood plasma samples from the dying person and prepare to inject vaccines to build immunity and be done with it.

Even both WHO and CDC lack the support of a dedicated *Medical Military* ready to be activated by a push of a button. And our governments are running helter-skelter for ideas, innovators and implementors. Irrespective that everyone in the government floats and boasts in a virtual database of ideas. To put it wildly, guys, we are in the thick danger of extinction. Don't drive us further into it.

We could learn from the procedures in Passive Air Defense. Incorporate sirens that send people into ditches or bunkers in minutes, mobile units recalled rapidly, and reserves of pensioned or retired army, paramilitary or medical doctors marshalled Army like logistics support to be activated. And above all, solid support for research and development should lead all efforts.

But ... By June, there could be nearly twice as many coronavirus deaths in the U.S. and eight times as many cases reported as there are today, according to projections from an internal Trump administration document that was obtained by The New York Times in May-2020.

Currently, the country's daily death toll stands around 1,750; but by June 1, it could reach 3,000 daily deaths, according to the projections based on models.

## 5. **Phantom of the Opera**

I am reminded of the theme song of a great musical play 'Phantom of the Opera' as I think about the virus hiding inside our bodies.

[CHRISTINE] In sleep he sang to me, in dreams he came
That voice which calls to me and speaks my name
And do I dream again? For now, I find
The Phantom of the Opera is there, **inside my mind**

[PHANTOM] Sing once again with me our strange duet
My power over you grows stronger yet
And though you turn from me to glance behind
The Phantom of the Opera is there, **inside your mind**

## 6. **Our intriguing human body**

Let's see how much we know about our bodies, as that should be part of our education. Our body has sufficient protection mechanisms inserted by nature yet it remains a vulnerable target.

The author is not qualified to say much about life's biological perspective but was recently forced to quickly study it. Corona became our utmost concern during the last few months, and it may remain so forever, so better understand our bodies a little. The information below is just a telescopic explanation mainly to appreciate the other chapters in the book.

If you find it too technical just ignore the chapter and jump to the next. All efforts have been made to keep the headings in an easily understandable sequence.

**Life and evolution**

The biologists have a simple way of defining life. Life continues while you or any organ of yours can move around using internal forces, and death terminates this contract. It's a *cell* that keeps us alive, and you know what, I came to know as late as 17-Apr-2020 from an official report, that even if we are dead the cells might keep operating inside the dead body.

The cell is the basic structural and functional unit of all known living organisms. So naturally, it becomes the building block of our life too and a messenger of death if fails in its job.

9

---

[9] A forensic practitioner working in Bangkok, Thailand, most likely caught the virus from a deceased patient, according to the report, which was posted online 2020 April 11 as a preprint for the Journal of Forensic and Legal Medicine.

The evolutionary history of life is something that Charles Darwin [10] laid down in his theory of biological evolution named *Origin of Species*. He travelled in far reaches of the South American Islands and found fossils of strange animals, no longer visible now. He arranged his acquired knowledge in that book and since then different people quote its contents in diverse contexts, shapes, or forms.

It is a theory that states, 'all species of organisms arise and develop through the natural selection of small, inherited *variations* that increase the individual's ability to compete, survive, and reproduce. It seems like a statement alone, not easy to comprehend though, but it's a theory that proves why we humans exist on the planet today.

If someone told me that our ancestor was an ancient fish called *Tiktaalik,* that lived 375 million years ago, at least I for one wouldn't believe. Understandably this fish had shoulders, elbows, legs, wrists, a neck, and many other basic parts that eventually became part of us.

Nature or God has configured our bodies so admirably that until now you cannot find a man who can describe it comprehensively. Different classes of people, scientists of varying disciplines particularly from the medical profession or the religious preachers laden with several philosophies are groping in the dark, making conjectures about this reality of life.

**Wonderous inside of our body**

**Cells**

Almost 99% mass of the human body is made up of six *elements*: oxygen, carbon, hydrogen, nitrogen, calcium, and phosphorus. Only about 0.85% is composed of another five elements: potassium, sulphur, sodium, chlorine, and magnesium.

---

[10] English naturalist **Charles Robert Darwin** (1809–1882)

All eleven make up our life and we must remain amazed at how the whole machinery continues to operate every second in billions of seconds of our life, unendingly. And when it stops, we die. Failure of the oscillator in the heart's natural pacemaker is one known cause of our death, but what if our cells stop to replenish the DNA chains! Seems it will make an eerie plot of a mystery book.

The human body is composed of 32.7 trillion *cells*. They provide structure for the body, take in nutrients from food, convert them into energy, and carry out specialized functions. The terms nutrient, food, energy, etc are our own coined terms and might mean different things to people. *God has nothing to do with it though.*

It's quite surprising that people adore vegetarian food and drastically oppose the non-vegetarian diet. Why? Is it simply a myth that a particular food is complete or incomplete with its proteins, or it's just an argument to be won through? Whether or not you are a vegetarian the body generates adequate energy to survive and to continue living.

I have seen under-nourished bodies survive with or without insignificant food for days. I have survived without eating an egg for 72 years. It's quite strange that people eat chicken with a bubbling gusto but twitch their noses at the sight of a cooked dog, frog, reptile, or a snake.

I will not recommend you any type of food, because I think your body contains the programmed *genes* and *proteins* inside, from your birth time, and you could conveniently use this program for your daily diet. And perhaps you know better which diet jells with your body.

This is precisely the reason that some expensive nutrition experts advise you on maintaining your initial (childhood) diet since your body was formed with those guts and with those bacteria. No alien diet like outlandish salads, cheeses, or soups will shed your weight if you are an Indian. You can shut your eyes off to those good-for-nothing mutterings.

Still, I think we might be ignorant about what is right and what could be wrong. Going by Darwin's theory, all weak species are likely to be superseded by the strong ones. So, in all sanity humans may kill animals and powerful animals may find their food in humans.

Who knows, a new dinosaur alike is breeding up in the thick interiors of rain forests in Amazon delta or icy holes deep in Antarctica? Survival of the fittest, though in a billion years hence.

## Cell Types

Over 200 different cell types form up the human body. Each type carries out a particular function, either solely, but usually by forming a particular *tissue*. Different tissues then combine to form specific *organs* or *organelles* (small organs) The organs work like a factory where every type of cell performs its job.

Some examples of cells are

- Stem cells
- Red blood cells
- White blood cells (Neutrophils, Eosinophils, Basophils, Lymphocytes)
- Platelets
- Nerve cells - These cells are specialized for communication
- Neuroglial cells
- Muscle cells - These cells are specialized for contraction (Skeletal muscle cells, Cardiac muscle cells, Smooth muscle cells)
- Cartilage cells

The longest cell is the nerve cell. The largest is the female ovum. The smallest cell is male sperm.

According to Wikipedia, between 50 and 70 billion cells die and re-emerge revitalized every ***day*** in an average adult. Isn't there another universe inside of us!

All cells have a cell membrane, cytoplasm (a jelly-like fluid), and genetic material contained in the cell *nucleus* (also called the nucleus *genome*). Some genetic material lies in the mitochondria (the mitochondrial *genome*) too.

**Eukaryotic and Prokaryotic cells**

Two types of cells exist in our bodies.

Eukaryotic cells are the ones that have organelles to house the nucleus and other items. These cells dwell in plants and animals. Being a bit more advanced and complex have a possibility of superseding our human existence!

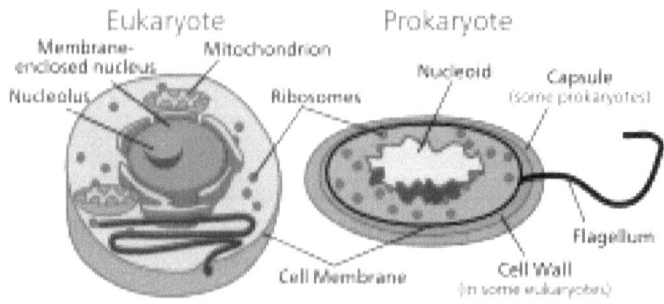

Prokaryotic cells, the other type, don't have a nucleus or membrane. They do have genetic material but it is not inside any nucleus.

**The fascinating inside of a cell**

*Genomes*

Dictionary suggests the name Genome, is a blend of the words, *gene,* and *chromosome*. Genome determines the physical characteristics of a person.

The human genome is mostly the same in all people. But there are variations across the genome domain that accounts for about 0.001

per cent of each person's DNA [11] and contributes to differences in appearance and health. Closely related people have more similar DNA.

Each genome contains all information needed to build and maintain or retain that organism. For this purpose, a copy of the entire genome is contained in all cells that have a nucleus.

*Genes*

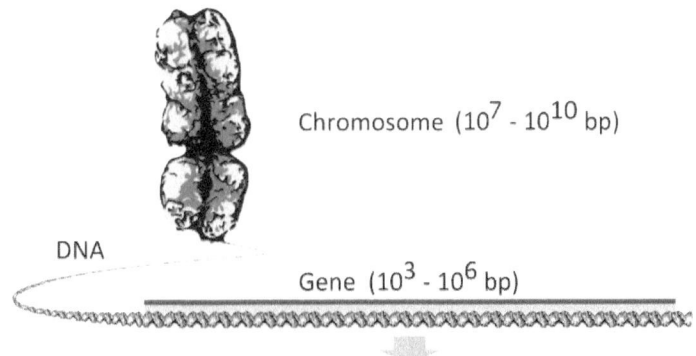

Chromosome ($10^7$ - $10^{10}$ bp)

DNA

Gene ($10^3$ - $10^6$ bp)

Function

A gene is the basic physical and functional unit of *heredity*. It is a sequence of *nucleotides* in DNA or RNA, [12] that encodes the synthesis of a gene product, which is either RNA or protein. Some genes act as instructions or genetic codes to make the protein molecules. Others produce only RNA.

*Chromosome*

Ah, here it is. The most important aspect of human life.

A chromosome is a DNA molecule that contains a part or all of the genetic material of an organism. Most eukaryotic chromosomes include packaging proteins that bind or condense the DNA molecule.

---

[11] DNA (Deoxy Ribonucleic Acid)
[12] RNA (Ribonucleic acid)

These packaging proteins are aided by chaperone proteins to prevent entanglement of the DNA.

The genome (genetic material) is divided into 23 pairs of linear *DNA* molecules or the chromosomes.

*Chromatin*

Within chromosomes, DNA is held in complexes with structural proteins called *histones*. These proteins organize the DNA into a compact structure called *chromatin*. Its primary function is packaging long DNA molecules into more compact, denser structures.

*Histones*

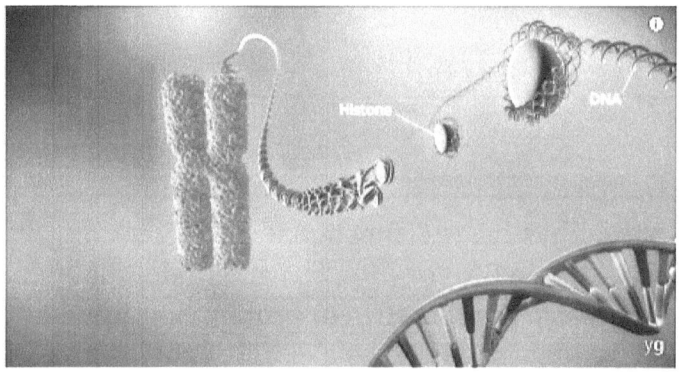

*Chromosome – male or female child?*

*Male sperm Cell*

Not that we have any control unless we fiddle with things, yet wisdom is knowing what nature has provided for us. Chromosomes are our mysterious treasures as far as offspring production is concerned.

*Sex chromosomes*

Humans have two sex chromosomes, the X and Y. They form the *23rd pair* of the 23 pairs of chromosomes in each cell. Females have two copies of the X chromosome, while males have one X and one Y chromosome. All *egg* cells (that means of a female) contain an X chromosome, while *sperm* cells of a man contain an X or a Y chromosome.

## SPERM CELL

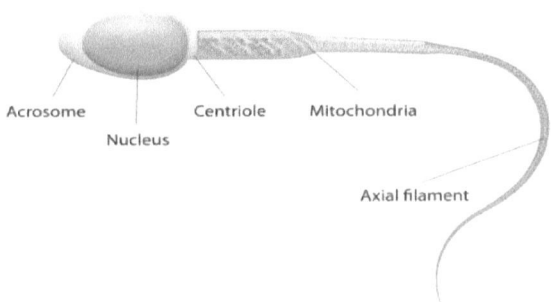

A man's testicles constantly produce new sperm in spermatogenesis. The full process takes about 64 days. During spermatogenesis, testicles make about 1,500 per second amounting to several million sperm per day. By the end of a full sperm production cycle, one can regenerate up to 8 billion sperms. Not bad really.

A man may ejaculate 40 million to 150 million sperms, which start swimming upstream toward the fallopian tubes, on their mission to fertilize an egg. A fast-swimming sperm can reach the egg in half an hour, while others may take days. The male sperm can live up to 48-72 hours inside a woman's body.

To get a boy offspring, the male sperm with the Y chromosome should be deposited as close as possible to the egg. The Y chromosome can't do it since it has a short lifespan and cannot live longer than 24 hours.

If a man has sex even a few days before his female partner ovulates, there's a chance that she may get pregnant. The Y chromosome, which makes boys, contains less DNA than the X chromosome for girls, this is why a Y chromosome swims faster in viscous liquids.

*The male X chromosome combines with the mother's X chromosome to make a baby girl (XX) and the male's Y chromosome will combine with the mothers to make a boy (XY).*

### Gonads (Ovaries and Testes)

The female *gonad*, the *ovary* or 'egg sac', is one of a pair of reproductive glands in women. They are located in the pelvis, one on each side of the uterus. Each ovary is about the size and shape of an almond. The ovaries have two functions, they produce eggs (ova) and the female hormones.

In males, the gonads are called *testes*.

### Sex rendering in the foetus

During early development the gonads the foetus remains undifferentiated; that is, all foetal genitalia are the same and are phenotypically *female*. After approximately 6 to 7 weeks of gestation, the expression of a gene on the Y chromosome induces changes that result in the development of the testes.

### Genetic Code

The genetic code is the set of rules used by cells to translate information encoded within genetic material (DNA or mRNA sequences of nucleotide triplets, or codons) into **proteins**.

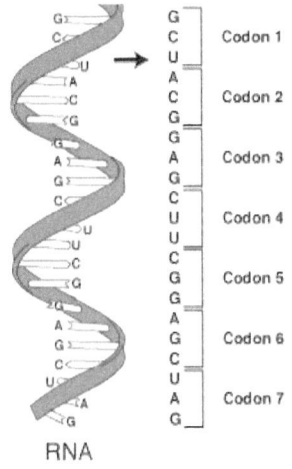

RNA

| | Codon 1 |
| Codon 2 |
| Codon 3 |
| Codon 4 |
| Codon 5 |
| Codon 6 |
| Codon 7 |

The translation is accomplished by the *ribosome*, which links amino acids in an order specified by messenger RNA (mRNA).

The nucleotides are abbreviated with the letters A, U, G, and C. This is mRNA, which uses U (Uracil). DNA uses T (thymine) instead.

This mRNA molecule will instruct a ribosome to synthesize a protein according to this code.

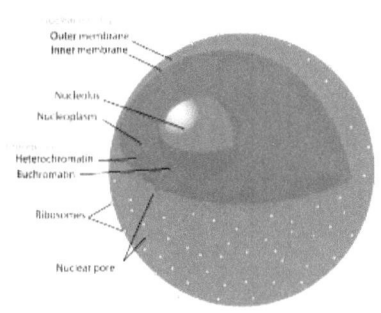

*Nucleolus*

The nucleolus is the largest structure in the nucleus of eukaryotic cells. It is best known as the site of ribosome biogenesis.

Nucleoli (plural) are made of proteins, DNA, and RNA.

*Ribosomes*

Ribosomes are macromolecular machines that perform biological protein synthesis (mRNA translation). Ribosomes link amino acids together in the order specified by the codons of messenger RNA (mRNA) molecules to form a polypeptide.

*Cytoplasm*

The cytoplasm is a thick solution that fills each cell and is enclosed by the cell membrane. It is mainly composed of water, salts, and proteins. The cytoplasm includes all of the material inside the cell and outside of the nucleus.

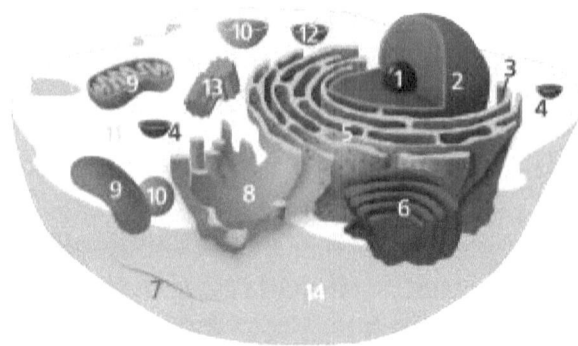

1. Nucleolus
2. Nucleus
3. Ribosome (little dots)
4. Vesicle
5. Rough endoplasmic reticulum
6. Golgi apparatus (or "Golgi body")
7. Cytoskeleton
8. Smooth endoplasmic reticulum
9. Mitochondrion
10. Vacuole
11. Cytosol (the fluid that contains organelles, comprising the cytoplasm)
12. Lysosome
13. Centrosome
14. Cell membrane

Cytoplasm helps the ribosomes to float around in and around its fluid/jelly and while they are floating around, they get attached to a *rough endoplasmic reticulum rER*.

The endoplasm reticulum is a membrane-enclosed passageway for proteins synthesized by the ribosome.

The synthesized proteins emerge from the endoplasm through *vesicles* (seen as polka dots in the figure on page 6-37). A vesicle is a structure within or outside a cell, consisting of liquid or cytoplasm enclosed by a lipid [13] bilayer.

*Golgi apparatus* is a complex of vesicles and folded membranes within the cytoplasm that is involved in secretion and it becomes an intracellular transport. As proteins move through the Golgi apparatus, they are customized by folding the protein in useable shape or adding other materials such as lipid or carbohydrates, etc.

*Nucleic acids*

Nucleic acids are composed of monomers, which are called *nucleotides.* Three components are important in this composition: a 5-carbon sugar, a phosphate group, and a nitrogenous base. A simple *ribose* makes the RNA polymer (*ribonucleic* acid); and if the sugar is derived from ribose as *deoxyribose*, the polymer is DNA (*deoxyribonucleic* acid).

Their function is to transmit and express the genetic information from the cell nucleus - to the interior operations of the cell and ultimately to the next generation of each living organism.

The biological information contained in an organism is encoded in its DNA and RNA sequences. Most organisms use DNA for their long-term information storage, but some *viruses* have RNA as their genetic material.

A genome sequence is the complete list of the nucleotides (A, C, G, and T for DNA genomes) that makeup all the chromosomes of a species.

---

[13] **Lipid** is an oily organic compound insoluble in water but soluble in organic solvents. It can be used for checking cholesterol and triglycerides. An excess amount of blood lipids can cause fat deposits in your artery walls, increasing your risk for heart disease.

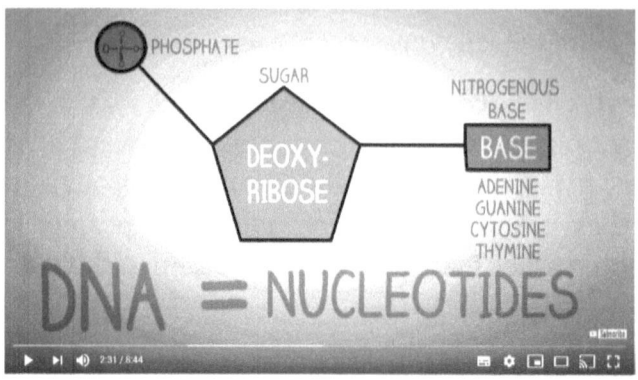

DNA- Structure and function of Deoxyribonucleic Acid (DNA)

Those of you who are educated on organic chemistry know that a polymer is a chemical compound with molecules bonded together in long, repeating chains. Polymers are both man-made and natural. Because of their structure, polymers have unique properties that can be tailored for different uses. For example, natural polymers (also called biopolymers) include silk, rubber, cellulose, wool, amber, keratin, collagen, starch, DNA, and shellac.

### DNA (Deoxy Ribonucleic Acid)

For most of its life cycle DNA stays inside the nucleus of the cell in the form of a coil, storing and coding all the genetic information.

Strings of nucleotides are bonded to form helical *backbones* - typically, one for RNA, two for DNA - and assembled into chains of *base-pairs* selected from the five primary, or canonical nucleobases, which are: Adenine, Cytosine, Guanine, Thymine, and Uracil. Please note thymine occurs only in DNA and uracil only in RNA.

Each is referred generally by A, G, C, T, and which are represented by the steps or rungs of the ladder. Generally, A, T are bonded together by a hydrogen bond. Likewise, C, G is another bonded pair. These are also called nitrogen bonds and these bonds give the DNA the strength and keep it stable.

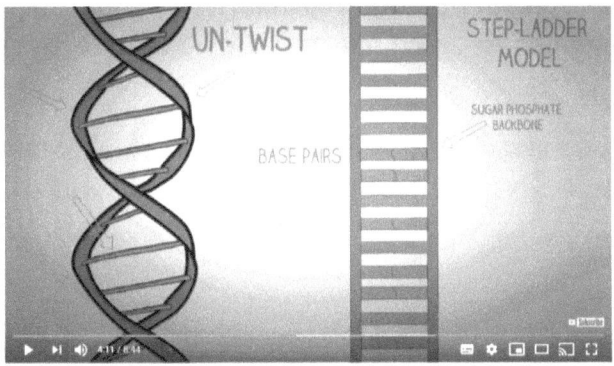

DNA: Structure and function of Deoxyribonucleic Acid (DNA)

DNA serves two important cellular functions. It's the genetic material that passes from parent to offspring and it serves to construct the proteins necessary for the cell to perform all of its functions. During gene expression, the DNA is first copied into RNA.

### RNA (Ribonucleic acid)

RNA has various biological roles in coding, decoding, regulation, and expression of genes. RNA and DNA are nucleic acids and along with lipid, proteins and carbohydrates constitute the four major macromolecules essential for all known forms of life.

### Proteins

Proteins are large biomolecules, or macromolecules, consisting of one or more long chains of amino acid residues joined together by peptide bonds. These bonds are broken by adding water.

Proteins perform catalysing metabolic reactions such as DNA replication, responding to stimuli, providing structure to cells, and organisms, and transporting molecules from one location to another.

6-36

Proteins differ from one another primarily in their sequence of amino acids, which is dictated by the nucleotide sequence of their genes

Many proteins are *enzymes* and others have structural or mechanical functions such as forming scaffoldings for protection of the cell shape.

*Enzymes*

Enzymes are proteins that *catalyse* or increase the rate of (bio) chemical processes.

We know enzymes from the cooking practices of making bread using yeast, and fermenting sugar into alcohol. Not only enzymes are selective and specific to types of chemical reactions, ie one enzyme does not help more than one type of reaction, but also reactions simply cannot take place without the presence of the enzymes.

One must learn how to use enzymes whenever your stomach is hard and full. Enzymes help digestion.

**Growth and metabolism**

Between successive cell divisions, cells grow through the functioning of cellular metabolism which is a process by which individual cells process nutrient molecules.

Metabolism has two distinct divisions: Catabolism, in which the cell breaks down the complex molecules to produce energy, and anabolism the opposite, in which the cell uses energy and the reducing power to construct complex molecules to perform other biological functions.

Complex sugars consumed by the organism can be broken down into simpler sugar molecules called glucose.

DNA dies every day and new replicates are created every day.

When a cell is ready to divide and proliferate into its copies, the DNA condenses into chromosomes. When a cell divides, it must replicate the DNA in its genome so that the two daughter cells have the same genetic information as to their parent. The double-stranded structure of DNA provides a simple

DNA is made up of a double helix of two complementary strands. During replication, these strands are separated and then each strand's complementary DNA sequence is recreated by an enzyme called DNA *polymerase* and bonded onto the original strand. Each strand of the original DNA molecule then serves as a template for the production of its counterpart, a process referred to as *semiconservative replication*.

Cellular proofreading and error-checking mechanisms ensure near-perfect fidelity for DNA replication for its daughters.

Enormous efforts have gone into the development of experimental methods to determine the nucleotide sequence of biological DNA and RNA molecules, and today hundreds of millions of nucleotides are sequenced daily at genome centres and smaller laboratories worldwide

**Amino Acids**

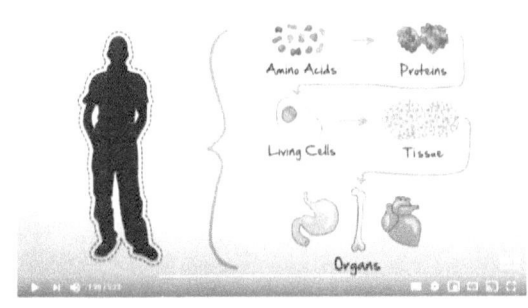

Amino acids are organic compounds that contain amine ($-NH_2$) and carboxyl ($-COOH$) functional groups, along with a side chain (R group) specific to each amino acid.

**Bacteria (living organisms)**

Bacteria are one-celled organisms that sometimes cause infectious diseases but, very often, are harmless. They are diverse and come in three shapes, a sphere, spiral, or rod. And they thrive in many different types of environments, extremes of cold or heat.

Bacteria do not have a membrane-bound nucleus, and their genetic material is typically a single circular bacterial chromosome of DNA located in the cytoplasm in an irregularly shaped body called the nucleoid.

Some of them make their home in people's intestines, where they help primarily to digest food.

Some bacterial infections are contagious, meaning that they can be spread from person to person. There are many ways this can occur, including:

- close contact with a person who has a bacterial infection, including touching and kissing.
- contact with the body fluids of an infected person, particularly after sexual exposure, or when an infected person coughs or sneezes
- transmission from mother to child during pregnancy or birth
- coming into contact with contaminated surfaces, such as doorknobs or faucet handles, and then touching your face, nose, or mouth

In addition to spreading from person to person, bacterial infections can also be spread through the bite of an infected insect, or by consuming contaminated food and water. Only a handful of bacteria known as *pathogenic* bacteria cause infections in humans. Some examples of bacterial infections are:

- strep throat
- urinary tract infection (UTI)
- bacterial food poisoning
- gonorrhoea
- tuberculosis
- bacterial meningitis

- cellulitis
- Lyme disease
- tetanus

Inappropriate use of antibiotics is known to have caused bacterial diseases that are resistant to treatment.

## Pathogens

Are disease-producing agents, especially viruses and certain kinds of bacteria.

## The human immune system

The human immune system has a significant role in our life. There are a few ways in which a human being can die.

- Non-communicable diseases such as cardiovascular disease, cancers, diabetes, and respiratory diseases
- Natural disasters that are visibly uncontrollable by humans such as viruses, and plagues of the world.
- Deaths due to injuries in road accidents, homicides, drowning, fire-related accidents, floods, and suicides

Our immune system takes care of the first two types.

The system is a collection of structures and processes within the body to protect against disease or other potentially damaging foreign bodies. When functioning properly, the immune system identifies a variety of threats, including viruses, bacteria, and parasites, and distinguishes them from the body's healthy tissue.

The type of *innate* immunity is the immune system you're born with, and it mainly consists of barriers on and in the body that keep foreign threats out. Components of innate immunity include skin, stomach acid, enzymes found in tears and skin oils, mucus, and the cough reflex. Innate immunity is non-specific, meaning it *doesn't* protect against any specific threats.

*Adaptive*, or acquired, immunity is another type of immune system. It targets specific threats to the body. In adaptive immunity, the threat must be processed and recognized by the body, and then the immune system creates antibodies specifically designed to the threat. After the threat is neutralized, the adaptive immune system *remembers* it, which makes the future responses to the same germ, more efficient.

**Lymph nodes -** Are small, bean-shaped structures that produce and store cells that fight infection and disease. These are part of the lymphatic system - which otherwise consists of bone marrow, spleen, thymus, and lymph nodes. Lymph nodes also contain lymph, the clear fluid that carries those to different parts of the body. When the body is fighting infection, lymph nodes can become enlarged and feel sore.

**Spleen -** The largest lymphatic organ in the body, which is on your left side under your ribs and above your stomach, contains white blood cells that fight infection or disease. Spleen also helps control the amount of blood in the body and disposes of old or damaged blood cells.

**Bone marrow -** The yellow tissue in the centre of the bones produces white blood cells. This spongy tissue inside some bones, such as the hip and thigh bones, contains immature cells, called stem cells. Stem cells are prized for their flexibility in being able to morph into any human cell.

**Lymphocytes -** These small white blood cells defend the body against disease. B-cells Lymphocytes, make antibodies that attack bacteria and toxins, and T-cells, help destroy infected or cancerous cells.

**Thymus -** This small organ is where T-cells mature. This is situated beneath the breastbone (and is shaped like a thyme leaf, hence the name). It can trigger or maintain the production of antibodies that result in muscle weakness.

**Leukocytes -** These disease-fighting white blood cells identify and eliminate pathogens and are the second arm of the innate immune system. A high white blood cell count is referred to as leucocytosis.

*In the context of COVID-19 senior citizens beyond the age of 65 years have been cautioned to take care of their immune system.*

### Lifestyle going awry

A specific disease is important for all of us. Type 2 diabetes can slowly overpower your immune system unless you mend your ways with your lifestyle. It works like a slow poison degenerating your organs sporadically. So, it's better to keep a watch on the below-mentioned symptoms rather fastidiously.

Eating in moderation, sleeping well, and above all regular exercise are the best medications for your immune system recovery. And if I may, walking is better and perennial lasting than gym and yoga.

Signs you must observe very carefully in day to day life to know if diabetes has made her abode in your body. If you don't detect any one of these, diabetes will not hesitate to give you a warning. But it may be too late then.

- Fatigue
- Excessive hunger and thirst
- Frequent urination
- Weird smelling breath
- Yeast infections
- Frequent UTIs
- Erectile dysfunction
- Blurry vision
- Slow-healing sores
- Unexplained weight loss
- Nausea and vomiting
- Painful or numb feet and legs
- Swollen or tender gums

- Weird smelling breath
- Frequent UTIs
- Polycystic Ovary Syndrome
- Skin darkening
- Chronic dehydration
- Irritability/depression

## 7. **Disasters hit us hard** [14]

---

*Were the pandemics precisely man-made*

**N**atural disasters have something to do with bacteria, viruses, pathogens, and DNA compositions. And as the US president George Bush once remarked, 'these are killers that leave no one spared'.

We have a long history of disasters and we suffered immensely.

All previous epidemics/pandemics were global since they affected millions around the world. But the situation was different then. The population was scanty and means of travel and movement restricted. It's perhaps irrelevant to give the number of deaths in these disasters because our vista or perspective has changed drastically now.

For example, while we were looking at the world population of 300 Million in the year 500AD, and 1 billion in the year 1800AD when some listed (below) disasters occurred, today we are in the range of 7.8 Billion, a rise of at least 7.8 times more than 1800AD.

The world has been transformed into a global village, as people brag about, so are all its attributes ie viruses, diseases and wars. The rapidly flying jets cannot help but spread the viruses in the blink of an eye like no one ever thought.

So, by a conservative rough guesstimate, we could keep the losses thereof as *multiplied* by a factor of 7.8. Meaning that if 12 Million people died in the 3rd plague in 1855, on today's date it would have been about 100 Million. We must keep this perspective alive in our minds as we study the disasters now.

Further, the future transmittable viruses will be *radio-energetic* as passengers will carry them along much faster, needing the vaccine

---

[14] These texts have been taken from CDC, Google and Life Science

to be developed in matching time frames. Therefore, the future outbreaks will be dramatically devastating, infecting people much more than we've ever seen before.

It's terrible, shocking, and depressing.

## Disasters - They never left you in peace [15]

### 3000 BC – Circa - Prehistoric

About 5,000 years ago, an epidemic wiped out a prehistoric village in China. The bodies of the dead were stuffed inside a house that was later burnt away. The skeletons of juveniles, young adults, and middle-aged people were found inside the house. The archaeological site is now called *Hamin Mangha* and is one of the best-preserved prehistoric sites in north-eastern China.

Before the discovery of Hamin Mangha, another prehistoric mass burial that dates to roughly the same period was found at Miaozigou, in north-eastern China. Together, these discoveries suggest that an epidemic ravaged the entire region.

### 430 BC - Plague of Athens (100,000)

Around 430 BC an epidemic ravaged the people of Athens and lasted five years. Some estimates put the death toll as high as 100,000 people.

The Greek historian Thucydides (460-400 BC) wrote that 'people in good health were all of a sudden attacked by violent heats in the head, and redness and inflammation in the eyes, the inward parts, such as the throat or tongue, becoming bloody and emitting an unnatural and fetid breath'.

---

[15] For further studies look at Google or Wikipedia

### 165 - Antonine Plague (5 M)

The Antonine Plague, which may have been smallpox brought to the Roman Empire by troops returning from campaigns. It might have killed over 5 million people in the Roman empire.

### 250-271 - Plague of Cyprian (5000 a day)

The Plague of Cyprian is estimated to have killed 5,000 people a day in Rome alone. In 2014, archaeologists in Luxor found what appears to be a mass burial site of plague victims. Their bodies were covered with a thick layer of lime (historically used as a disinfectant). The cause was suspected to be smallpox, pandemic influenza, and viral haemorrhagic fever.

### 541-542 - Plague of Justinian (10% world population died)

The Byzantine Empire was ravaged by the bubonic plague, which marked the start of its decline. The plague reoccurred periodically afterwards.

### 1346-1353 - The Black Death (2nd Plague or Bubonic Plague) (200 M)

This story started during the Byzantine Empire.

A flea-borne bacterial disease of rodents jumped to fleas and then to humans as a virus. The plague was caused by the bacteria *Yersinia Pestis*, which is still very much alive and generally seen in animal populations and transmitted by the bite of a flea.

Within three days, the patients vomited blood and died. Victorian scientists dubbed it the *Black Death*, the bubonic [16] plague. It earned its name from a symptom: lymph nodes that became blackened and swollen after bacteria entered through the skin.

Black Death, was so contagious and deadly that 'the mere touching of the clothes appeared to itself to communicate the malady to the

---

[16] an inflammatory swelling of a lymphatic gland, especially in the groin or armpit

toucher.' A healthy person who went to bed at night could be found dead in the morning!

On an average 5,000 people died every single day, and by the time we weathered the 'Black Death', as it was called, 100 million people died, which was close to half the population of Europe. In total, the plague may have reduced the world population from an estimated 475 million to 350–375 million in the 14th century.

It took 200 years for Europe's population to *recover* to its previous level, and some regions (such as Florence, a marvellous tourist spot in Italy) only recovered by the 19th century. Outbreaks of the plague recurred until the early 20th century.

The Italian Peninsula has been the epicentre of most of the epidemics probably because it is one of the most travelled countries. As is the case with coronavirus.

We now have the luxury of N95 masks and sanitizers to hoard (and for some people, a lot of toilet paper) but back then people had none and they turned to religion and *whipped* themselves to appease God's wrath.

Plague doctors wore a mask with a bird-like beak to protect them from being infected by deadly viruses, which they believed was *airborne*. To battle this imaginary threat, the long beak was packed with sweet smells, such as dried flowers, herbs, and spices.

Some of the cures they tried included: Rubbing onions, herbs, or a chopped-up snake (if available) on the boils or cutting up a pigeon and rubbing it over an infected body. Drinking vinegar, eating crushed minerals, arsenic, mercury, or even ten-year-old treacle!

What a pity.

**1520 - Measles, Smallpox, and Chickenpox (over 50M affected)**

Smallpox was an infectious disease caused by one of two virus variants, *Variola major* and *Variola minor*. The last naturally occurring case was diagnosed in October 1977, and WHO certified the global eradication of the disease in 1980. The risk of death following contracting the disease was about 30%, with higher rates among babies. Often those who survived had extensive scarring of their skin, and some were rendered blind.

The disease historically occurred in outbreaks.

In 18th-century Europe, it is estimated 400,000 people per year died from the disease, and one-third of the cases resulted in blindness. Smallpox is estimated to have killed up to 300 million people in the 20th century and around 500 million people in the last 100 years of its existence. As recently as 1967, 15 million cases occurred a year. Isn't it gloomy?

The initial symptoms of the disease included fever and vomiting. This was followed by the formation of sores in the mouth and a skin rash. Over several days the skin rash turned into characteristic fluid-filled bumps with a dent in the centre. The bumps then scabbed over and fell off, leaving scars.

Transmission occurred through inhalation of airborne Variola virus, usually, droplets expressed from the oral, nasal, or pharyngeal mucosa of an infected person. It was transmitted from one person to another primarily through prolonged face-to-face contact with an infected person, usually within a distance of 1.8 m (6 feet), but could also be spread through direct contact with infected bodily fluids or contaminated objects (fomites) such as bedding or clothing.

Rarely, smallpox was spread by a virus carried in the air in enclosed settings such as buildings, buses, and trains.

The term smallpox was first used in Britain in the early 16th century to distinguish the disease from syphilis, which was then known as the *great* pox. Smallpox was highly contagious, but generally spread more slowly and less widely than some other viral diseases, perhaps because transmission required close contact and occurred after the onset of the rash.

The overall rate of infection was also affected by the short duration of the infectious stage. In temperate areas, the number of smallpox infections was highest during the winter and spring.

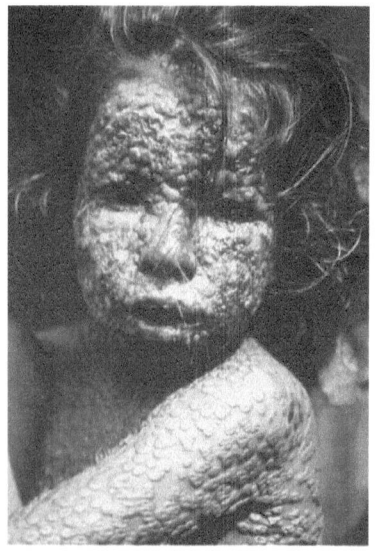

In tropical areas, seasonal variation was less evident and the disease was present throughout the year.

**1545-1548 - Cocoliztli epidemic (15 M)**

The infection that caused the cocoliztli epidemic was a form of viral haemorrhagic fever that killed 15 million inhabitants of Mexico and Central America.

17

---

17 Picture from CDC

## 16th Century - American Plagues (90% of the indigenous population)

The American Plagues are a cluster of Eurasian diseases brought to the Americas by European explorers. These illnesses, including smallpox, contributed to the *collapse* of the Inca and Aztec civilizations. Some estimates suggest that 90% of the indigenous population in the Western Hemisphere was killed off.

## 1665-1666 Great Plague of London (100,000)

The Black Death's last major outbreak in Great Britain caused a mass exodus from London, led by King Charles II. The plague started in April 1665 and spread rapidly through the hot summer months. Fleas from plague-infected rodents were one of the main causes of transmission. By the time the plague ended, about 100,000 people, including 15% of the population of London, had died.

But this was not the end of that city's suffering. On Sept. 2, 1666, the Great Fire of London started, lasting for four days and burning down a large portion of the city.

## 1855 - 3rd plague (12M affected)

The third plague was a major bubonic pandemic that began in Yunnan, China, in 1855. This plague spread to all inhabited continents and ultimately led to more than 10 million deaths in India and 2 million in China. According to the WHO, the pandemic was considered active until 1960, when worldwide casualties dropped to 200 per year.

The *'third'* word implies that it was the third major bubonic plague outbreak to affect European society. The first was the Plague of Justinian, which ravaged the Byzantine Empire in 541 and 542. The second, the Black Death, which killed at least one-third of Europe's population in a series of expanding waves of infection from 1346 to 1353.

Casualty patterns indicate that waves of a pandemic may have come from two different sources. The first was primarily bubonic and was carried around the world through ocean-going trade, through transporting infected persons, rats, and cargoes harbouring fleas. The second, more virulent strain, was primarily pneumonic with a strong person-to-person contagion.

### 1918-1919 - The Spanish Flu (50 M)

While it's likely that the 'Spanish Flu' originated in Spain due to its name, scientists are still unsure of its source. France, China, and Britain have all been suggested as the potential birthplace, as is the United States wherein at a military base in Kansas on March 11, 1918, the first known case was reported.

More recently, experts have proposed another hypothesis. The Spanish flu originated somewhere in northern China in late 1917 and moved to western Europe where the French and British governments recruited to perform manual labor, to free up troops for wartime duty. A total of 140,000 Chinese labourers were moved.

It was a strain of H1N1 that ran rampant after World War 1(which again materialized as Swine Flu in 2009). It killed 50 million worldwide, that's more than all of the soldiers and civilians killed during World War I combined. The pandemic's grasp stretched from the United States and Europe to the remote reaches of Greenland and the Pacific islands.

Its victims included the likes of President Woodrow Wilson, who contracted it while negotiating the Treaty of Versailles in early 1919.

Much like it has been the trend, the world handled it very poorly. Officials downplayed the threats, and as people continued to perish, confidence in governments was lost. At that time, there were no effective drugs nor vaccines to treat this killer. Citizens were ordered to wear masks, schools, theatres and businesses were shut and bodies piled up in makeshift morgues before the virus ended its deadly global march. Was that enough?

## 1959-1961 - The Chinese Famine (30 M affected)

It is interesting to know how foolish humans disturbed nature's ecological balance and caused a catastrophe for themselves. China suffered two years of famine in late 1960, where close to 30 million people starved and died under the regime of Mao-Tse-Tung.

This crisis was caused because of an insipid initiative, 'The Great Sparrow Campaign' which was also called, 'The Great Leap Forward'. The idea originated from the premise that the heavy population in China could convert its agricultural economy by exporting food grains to a country that was industry-based but might need food grains. This campaign was expected to upgrade China at par with the heavy economies of the world.

For this, it was necessary to shoot down sparrows that consumed substantial food grains and it was hoped that such an effort would increase the overall effective agricultural produce. Due to the sustained efforts of people and the army, millions of sparrows were chased systematically and killed. People wore garlands of dead

sparrows to celebrate their victory over nature. Not a moment did they realize that in the process they killed all the insects that the sparrows eat.

The shorthorn grasshopper (*Tiddi*) out of them returned the compliment by attacking and eating away the increased vegetable produce. It's most unfortunate fall out was the country was submerged into a great famine.

Foods became so scarce that the regime loosened its control of what qualified as *consumable* and *farmable* meat, breaking the moderation of livestock and extending it to the unconventional animals, such as *bats, pangolins, turtles, and snakes, among others.*

**1981 – 0nwards - HIV (1 M continues even now)**

HIV, a virus that causes AIDS [18], was probably originated from a chimpanzee and was transferred to humans in West Africa in the 1920s. The virus made its way around the world and AIDS became a pandemic by the late 20th century. Now, about 64% of the estimated 40 million living with HIV live in sub-Saharan Africa.

No longer mysterious to people, HIV has something to do with sex or needles. It is a virus spread through certain body fluids like blood, semen, vaginal and rectal fluids, and breast milk, that attack the body's immune system, specifically the CD4 cells, often called T cells. Over time, HIV can destroy so many of these cells that the body can't fight off infections and disease.

The virus doesn't spread in air or water, or through casual contact. It's estimated that over a million Americans are currently living with HIV. Of those, 1 in 5 don't even know they are. If AIDS develops further it makes the person vulnerable to a wide range of illnesses, including:

- pneumonia
- tuberculosis
- oral thrush, a fungal infection in the mouth or throat
- cytomegalovirus (CMV), a type of herpes virus
- cryptococcal meningitis, a fungal infection in the brain
- toxoplasmosis, a brain infection caused by a parasite

---

[18] acquired immunodeficiency syndrome

- cryptosporidiosis, an infection caused by an intestinal parasite
- cancer, including Kaposi's sarcoma (KS) and lymphoma

For decades, the disease had no known cure, but medication developed in the 1990s now allows people with the disease to live a normal life span with regular treatment. Even more encouraging, two people have been cured of HIV as of early 2020.

## 2009-2010 - H1N1 Swine Flu (600,000 in one year)

The 2009 swine flu pandemic was caused by a new strain of H1N1 that originated in Mexico in the spring of 2009 before spreading to the rest of the world. In one year, the virus-infected as many as 1.4 million people across the globe and killed between 151,700 and 575,400 people, according to the CDC.

## 2014-2016 - West African Ebola

The first known cases of Ebola occurred in Sudan and the Democratic Republic of Congo in 1976, and the virus may have originated in bats.

Ebola ravaged West Africa between 2014 and 2016, with 11,325 deaths. The first case to be reported was in Guinea in December 2013, then the disease quickly spread to Liberia and Sierra Leone.

So far, there is no cure for Ebola, although efforts at finding a vaccine are on. Reports indicate that it started again in 2020.

Commentary on disasters

Most of the pandemics are caused by viruses that generally arrive from rats, bats, rodents, pigs, camels, and other animals. These are also characterized by the failure of people to find a vaccine quickly, in order words not well in time to save the people.

One observation can be made though. Most of the viruses had a source which people could not determine or determined rather dubiously. We can ignore these inadequacies because our medical and science technology were not up to the mark as they are now.

We can deploy microscopes, light microscopes, nanoscopic probes, globally available virus databases and others to go after the viruses. If we cannot find the source deterministically, what use is then our science and technology?

As long as viruses go on taking birth, which they will, we need to augment our expert resources for making vaccines quickly. And as long as international flights remain in vogue, and they certainly will, we can't escape its wrath.

The purpose of including this chapter was to make you aware that such pandemics are not new and the world has coped with those despite heavy losses of lives.

# Timeline Pandemics

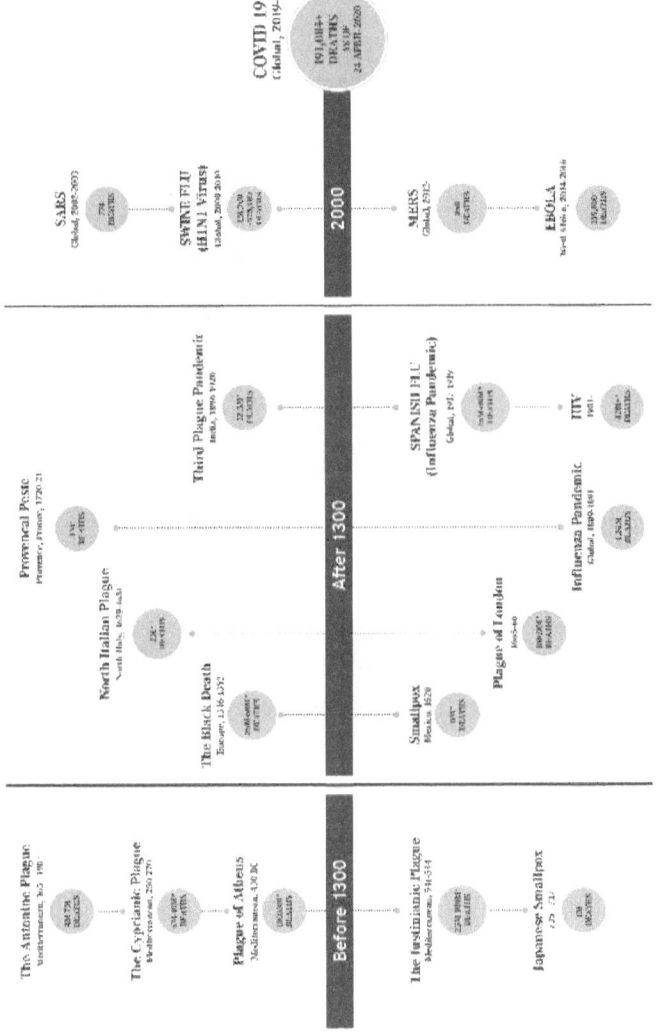

## 8. The Corona Virus Theories, Views and Opinions

This is a compilation of the chatter going around.

### Viruses – Infectious invaders

Viruses were discovered in 1892 and even in 2020 researchers are uncovering mysteries underlying these intruders.

Viruses are not quite living things and have no way to reproduce on their own but a sophisticated one like SARS-Cov-2 moulds its behaviour after the strengths and weaknesses of its hosts. It replicates its genetic material into host cells at the cost of their biological programming. Given the most common disease that follows transmission is a respiratory one, the lungs endure the worst of this process.

They have been described as *organisms at the edge of life* since they resemble organisms; in that, they possess genes, evolve by natural selection, and reproduce by creating multiple copies of themselves through self-assembly.

Sorrowfully and hopelessly some people's immune system lets them down and some viruses find their way inside their bodies. Here are some examples.

By the way, it doesn't matter which organ or organs of the body get affected by a virus, or what is shown on the symptoms radar, what matters is the fact that the virus is ultimately present inside the body.

Unless we find a way to remove, kill, or neutralise it all our techniques are futile. We can't make people survive its fury.

### Alzheimer

The theory that viruses might be playing a role in Alzheimer's disease received more support from a study published in June in the journal Neuron. Researchers looked at nearly 1,000 post-mortem brains from multiple brain banks, including brains from people *with and without* Alzheimer's disease.

Overview of
# Viral infections

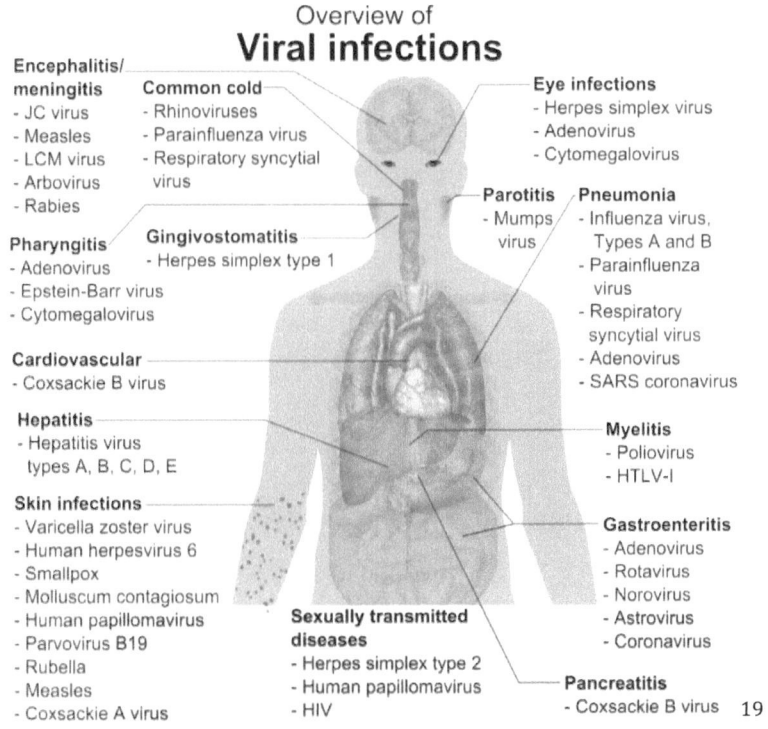

**Encephalitis/ meningitis**
- JC virus
- Measles
- LCM virus
- Arbovirus
- Rabies

**Common cold**
- Rhinoviruses
- Parainfluenza virus
- Respiratory syncytial virus

**Eye infections**
- Herpes simplex virus
- Adenovirus
- Cytomegalovirus

**Pharyngitis**
- Adenovirus
- Epstein-Barr virus
- Cytomegalovirus

**Gingivostomatitis**
- Herpes simplex type 1

**Parotitis**
- Mumps virus

**Pneumonia**
- Influenza virus, Types A and B
- Parainfluenza virus
- Respiratory syncytial virus
- Adenovirus
- SARS coronavirus

**Cardiovascular**
- Coxsackie B virus

**Hepatitis**
- Hepatitis virus types A, B, C, D, E

**Myelitis**
- Poliovirus
- HTLV-I

**Skin infections**
- Varicella zoster virus
- Human herpesvirus 6
- Smallpox
- Molluscum contagiosum
- Human papillomavirus
- Parvovirus B19
- Rubella
- Measles
- Coxsackie A virus

**Sexually transmitted diseases**
- Herpes simplex type 2
- Human papillomavirus
- HIV

**Gastroenteritis**
- Adenovirus
- Rotavirus
- Norovirus
- Astrovirus
- Coronavirus

**Pancreatitis**
- Coxsackie B virus  19

They sifted through genetic sequences taken from these brain tissues and identified which of the sequences were human and which were not. They found that the brains of deceased people with Alzheimer's disease had up to twice the level of two common strains of herpes viruses, compared with the non-Alzheimer's brains. The viruses could be part of the cause of the disease, or they could just speed up the progression of it.

---

[19] Picture from Wikipedia

## Herpes

Herpes simplex virus infections are common, with more than 80 per cent of the world's people infected with herpes simplex virus (HSV). The virus often remains in a *dormant* mode in the body, which is beneficial to people who are infected because the virus doesn't cause symptoms while dormant. However, it's also harder for the immune system to find and eliminate the virus while it is dormant.

In October 2017, researchers reported in the journal PLOS Pathogens that they had figured out how to induce the virus to enter its dormant mode. They also found the key proteins that are involved in waking it up.

The findings may have implications for treating or preventing herpes infections, the researchers said. The results could point towards ways to target certain viral proteins to prevent viruses from waking up, thus preventing symptoms and the spreading of the virus to other people, or could lead to ways to get the virus to remain *awake*, so that the immune system could eliminate it, the researchers said.

## Some lethal viruses

There have been many viruses on the planet so far, some deadly, some widely spread and some not yet combatted.

- 1918 – Spanish Influenza
- 1920 – Rabies – dogs and monkeys, it destroys the brain, it's a bad disease
- 1950 – Korea - Hantavirus – droppings of infected mice
- 1950 – Dengue - Philippines and Thailand – A mosquito bite
- 1967 - Marburg - Germany – From Monkeys
- 1976 – Ebola - Sudan and Congo – is spread through contact with blood or other body fluids, or tissue from infected people or animals
- 18Th Century – Smallpox - 380 million died

- 2002 - SARS-CoV - Guangdong province of southern China – through bats, nocturnal mammals
- 2012 - MERS-CoV – Saudi Arabia 2015 in South Korea - belongs to the same family of viruses as SARS-CoV and SARS-CoV-2. Likely originated in bats, infected camels before passing into humans.
- SARS-CoV-2 belongs to the same large family of viruses as SARS-CoV, known as coronaviruses

## Corona Viruses – rather syndromes

All coronaviruses sport spiky projections on their outer surfaces that resemble the points of a crown, or *corona* in Latin. These viruses have been responsible for several outbreaks around the world and in different *centuries.*

Coronaviruses (CoVs) primarily cause enzootic infections in birds and mammals but, in the last few decades, are infecting humans as well. The outbreak of severe acute respiratory syndrome (SARS) in 2003 and, more recently, Middle-East respiratory syndrome (MERS) has demonstrated the lethality of CoVs when they cross the species barrier and infect humans.

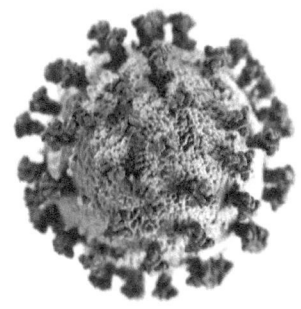

### The virus's spike proteins [20]

Most viruses have viral envelopes as their outer layer. Viruses which don't have membranes are termed *naked.*

In particular, the coronaviral genome encodes four major structural proteins: the *spike* (S) protein, *nucleocapsid* (N) protein,

[20] Molecular Biology and Virology Research Laboratory, Department of Medical Biosciences, University of the Western Cape, Cape Town, South Africa
**Dewald Schoeman & Burtram C. Fielding**

*membrane* (M) protein [21], and the *envelope* (E) protein, all of which are required to produce a structurally complete viral particle.

The CoV envelope (E) protein is a small, integral membrane protein involved in several aspects of the virus' life cycle, such as assembly, budding, envelope formation, and pathogenesis.

The protein coat (capsid) and the nucleic acid together are called the nucleocapsid. The entire intact virus is called the virion.

The capsid has three functions:

- it protects the nucleic acid from digestion by enzymes
- contains special sites on its surface that allow the virion to attach to a host cell
- provides proteins that enable the virion to penetrate the host cell membrane and, in some cases, to inject the infectious nucleic

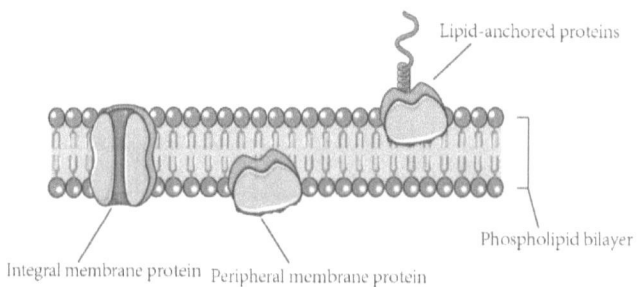

### Angiotensin-converting enzyme 2 (ACE2) Receptor

ACE2 is an enzyme attached to the outer surface of human cells in the lungs, arteries, heart, kidney, and intestines. Therefore, the virus

---

[21] Based on their structure, there are main three types of **membrane proteins**: the first one is **integral** membrane protein that is permanently anchored or part of the membrane, the second type is **peripheral** membrane protein that is only temporarily attached to the lipid bilayer or to other integral proteins, and the third one is lipid-anchored proteins.

affects those organs of the human body wherever the ACE2 detectors are found.

The virus, on the other hand, uses a special surface glycoprotein called a 'spike' *peplomer* [22] to connect to ACE2 and enter the host cell. The core of the virus contains genetic material that the virus can inject into vulnerable cells to infect them.

When the spike engages its ACE2 receptor on a host cell, it engulfs the receptor, resulting in the merger of the virus with the cell. This merger allows the virus to release its genetic material and hijack the cell's internal machinery. Once this happens, the virus sheds its coat and turns the cell into a factory that starts churning out new instances of the virus. A cascade is triggered.

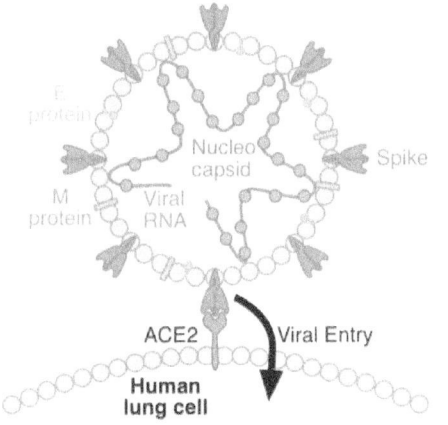

Shocking, it has done its job, now the ball is your court.

**Animal-based viruses**

Several coronaviruses utilize animals as their primary hosts and then evolve to humans. *Precursors* to both SARS and MERS coronaviruses appear in bats. The SARS virus jumped from *bats* to *civets* (small, nocturnal mammals) on its way into people, while MERS infected *camels* before spreading to humans. Evidence suggests that the novel coronavirus also jumped from bats to humans after passing through an intermediate carrier, although scientists have not yet identified the infectious middleman creature.

---

[22] A **peplomer** is a glycoprotein spike on a viral capsid or viral envelope. These protrusions bind only to certain receptors on the host cell. They are essential for both **host** specificity and **viral** infectivity.

## Human-based viruses

The four most common human coronaviruses - named *229E, NL63, OC43, and HKU1* – didn't jump from animals to humans but rather utilized humans as their natural hosts. These human-borne coronaviruses have presumably evolved to maximize *spread* amongst the population rather than *pathogenicity* to do so, meaning the viruses opted to maximize their own spread *rather than* harm their innocent human host using pathogens. This is quite different from the animal-based viruses.

There is also a thought that coronaviruses transmitted from animals to humans caused more-severe diseases, but the idea remains speculative.

## Mutations in viruses

How do the genes of the virus function? Viruses evolve by mutating (synonyms are metamorphosis, alteration, changes, transformations), which are the changes in their genetic code (genome sequence) over time. These occur randomly, and any changes that occur will be inherited by all copies of the next generation.

Based on current data, it seems as though **SARS-CoV-2's** genome is almost twice as large as the seasonal **flu** genome. So, it mutates more slowly than the seasonal flu, that is it has a mutation rate of fewer than 25 mutations per year, whereas the seasonal flu has a mutation rate of almost 50.

The seasonal flu, on the other hand, mutates roughly four times as fast as SARS-CoV-2. And this fact of quickness is precisely why flu can *evade* our vaccines,

The significantly slower mutation rate of SARS-CoV-2 gives us hope for an effective long-lasting vaccine.

**Differences between remedies – Viruses and Bacteria**

A vaccine is any preparation used as a preventive inoculation to confer immunity against a specific disease, usually employing an innocuous form of the disease agent, such as killed or weakened bacteria or viruses, to stimulate antibody production.

Antibiotics, however, only work against bacteria and other microorganisms. An antibiotic is any of a large group of chemical substances, as penicillin or streptomycin, produced by various microorganisms and fungi, having the capacity in dilute solutions to inhibit the growth of or to destroy bacteria and other microorganisms. These are used chiefly in the treatment of infectious diseases.

**The spread of Corona Virus**

There are two domains in which coronavirus spreads.

One, biologically it takes control of our body. It flows along with blood and air deeper and deeper into our organs and kills them softly and slowly. Sometimes fastly too. It is internal.

Second, it is not satisfied with one body and finds solace when it reaches out to other bodies in our contact. This domain can be called the people or the public domain.

It spread over people with an equal bias, whether you earn dollars or yuans, whether you are a Hindu or Christian, whether you are black or white. It has a panoramic view of the world. No holds barred.

To estimate how easily a virus spreads, scientists calculate its 'basic reproduction number,' or R0 (pronounced R-nought). R0 predicts the number of people who can catch a given bug from a single infected person. Currently, the R0 for SARS-CoV-2, the virus that causes the disease COVID-19, is estimated at 2.2, meaning a single infected person will infect about 2.2 others, on average. By comparison, the flu has an R0 of 1.3.

## Spread by touch

Before the virus was announced to the world, the Chinese had already hidden it for 6 weeks. We should give them the benefit of doubt. Maybe they were experimented with it and needed time. But the strange part is when the WHO supported the Chinese claim that COVID-19 *does not spread by touch*.

Very good. Seemed plausible and possible at that time.

But the whole world went into a frenzy when the cat was let out. The Chinese declaration came after a good 6 weeks and by that time the cat had become a ferocious lion. It jumped and attacked people. It didn't give them a chance. People couldn't take safeguards or precautions.

US Congressman James Comer along with members of the House Committee on Oversight and Reform wrote a letter to the Director-General of the World Health Organization requesting documents relating to their assistance in the Chinese government's coronavirus propaganda efforts.

In January 2020, the WHO had *promoted Chinese propaganda* on social media in claiming that the coronavirus *does not spread* by human transmission. Elsewhere, it is said to be contagious, meaning it *spreads* by touch.

## Contact transmission

With hindsight, we can now say that the virus surely spreads by contact.

Viral particles emitted from the respiratory tract of an infected individual land on a surface and when another person touches that surface, the virus travels to his hands. The hands then inadvertently touch the nose, mouth, or eyes and the virus sneaks into the body via the mucous membranes, thereby infecting the second person.

One study found that SARS-Cov-2 could remain viable on surfaces such as cardboard for up to 24 hours, and plastic and steel for 2 to 3 days.

[23] Joshua Santarpia and his colleagues found viral contamination in the air samples, on surfaces such as toilets, and frequently touched surfaces.

Also, on 26-Mar-2020, the CDC published a report on the coronavirus-stricken Diamond Princess cruise ship. An investigative team found traces of RNA from SARS-CoV-2 on surfaces throughout the cruise ship, in the cabins of both symptomatic and asymptomatic infected passengers, up to 17 days later — though no evidence suggests this viral RNA was still infectious.

(SARS-CoV-2 is an RNA virus, not DNA.)

Another CDC case report from Singapore also suggests contact with contaminated surfaces could transmit the virus.

In this case, a person who was infected with SARS-CoV-2, but not yet symptomatic, attended a church service. Later in the day, another person sat in the same seat. He also came down with COVID-19. Whether the virus was contracted via a contaminated surface, or potentially a lingering aerosol, couldn't be ascertained.

**Spread by physical proximity**

Some views though are gaining ground that Corona stays in the air and gets transmitted due to breathing nearby. Research Institute (NSRI) found traces of the virus in the rooms and outside in the hallways at the University of Nebraska Medical Centre, where confirmed COVID-19 patients were put in isolation.

---

[23] Leader of the group, **Joshua L Santarpia**, Danielle N Rivera, Vicki Herrera, M. Jane Morwitzer, Hannah Creager, George W. Santarpia, Kevin K Crown, David Brett-Major, Elizabeth Schnaubelt, M. Jana Broadhurst, James V. Lawler, St. Patrick Reid, John J. Lowe

And CDC also maintains an opinion that it might not be simply by touch. The presence of ill people in a space about *24 feet circle* is sufficient to throw infection. This is probably why we see a lot of fumigation being carried out using trucks and trailers, or with other machines in many countries.

## Spread by aerosol transmission

Researchers at Aalto University, the Finnish Meteorological Institute, the VTT Technical Research Centre of Finland, and the University of Helsinki used a supercomputer to simulate the spread of small viral particles leaving a person's respiratory tract through *coughing in a grocery store*. They simulated a scenario in which a person coughed in a store aisle between shelves, and took into account the ventilation around.

They found an aerosol 'cloud' spreads outside the immediate vicinity of the person coughing, and it dilutes as it expands. This person could cough and walk away but he leaves behind extremely small aerosol particles carrying the coronavirus. Although this process takes up to several minutes in the meantime if another person walks in, he could, in theory, inhale the small particles.

For the virus to be spread without being coughed or sneezed in large drops of mucus, it has to somehow be able to suspend in the air for long enough to infect passers-by.

A study published on 17-Mar in the New England Journal of Medicine stated that aerosolized virus particles could remain viable for up to 3 hours. What's not clear from this data how long the virus remains infectious in aerosols in *real-world settings*.

In that study, researchers had used an extremely high concentration of virus particles, which may not reflect those shed by people with the disease.

One case study is suggestive, however; a choir group in Washington met for a two-hour practice in early March. No one was symptomatic, so singers weren't coughing or sneezing out infected droplets. And everyone kept their distance. But when all was said and done, 45 people became infected with COVID-19 and at least two people died from the virus [24].

That suggested the viral particles were probably shed as aerosols by someone, before being inhaled or otherwise acquired by other choir members.

A 2019 study in the journal Nature Scientific Reports found that people emit more aerosol particles when talking and that louder speech volumes correlate to more aerosol particles being emitted.

**Direct or Indirect**

Findings indicate that disease might be spread through both direct (droplet and person-to-person) as well as indirect contact (contaminated objects and airborne transmission) and suggest airborne isolation precautions. Of course, we should wait for the WHO reports confirming this.

**Spread before or after hopping to humans**

A research group came up with two possible scenarios for the origin of SARS-CoV-2 in humans. One scenario follows the origin of other *similar* coronaviruses.

In that scenario, humans contracted the virus *directly* from an animal - civets in the case of SARS and camels in the case of the Middle East respiratory syndrome (MERS). In the case of SARS-CoV-2, that animal was a bat, which transmitted the virus to another

---

[24] Los Angeles Times reported a choir group in Washington in March 2020

intermediate carrier animal (possibly a pangolin, as some scientists have said). In that scenario, the genetic features that make the new coronavirus so effective at infecting human cells (its pathogenic powers) were in place *before* hopping to humans.

In the other scenario, those pathogenic features evolved only *after* the virus jumped from its animal host to humans. Some coronaviruses that originated in pangolins have a 'hook structure' (that receptor binding domain) similar to that of SARS-CoV-2 and passed its virus onto a human host. And post-injection the virus evolved to activate its other stealth feature - the cleavage site that lets it easily break into human cells.

**From dead bodies of COVID-19 victims**

A research report was published on 19-Mar-2020, about a forensic practitioner who died of the virus, marking the first case on record of a COVID-19 infection and death among medical personnel.

At the time of report just 272 people in Thailand - including a forensic practitioner and a nurse assistant - had tested positive for the new coronavirus. Most of these cases were imported, meaning they weren't from community spread. So, it's unlikely that the forensic practitioner caught the new coronavirus outside of work or even from a patient at the hospital, the researchers wrote.

There is a low probability of *forensic* professionals coming into contact with infected patients, but they might surely have contact with biological samples and corpses. The dead body of a recently deceased COVID-19 patient might be contagious, said Dr Otto Yang. [25]

'Absolutely, a dead body would be contagious at least for hours if not days,' Yang told Live Science in an email. 'The virus will still be in

---

[25] **Dr. Otto Yang**, a professor in the Department of Medicine and the Department of Microbiology, Immunology and Molecular Genetics at the David Geffen School of Medicine at UCLA

respiratory secretions, and potentially still be reproducing in cells that haven't yet died in the lungs.'

Therefore, COVID-19's longevity in the body can be problematic for people in the funerary industry.

For instance, following the reports from temples in Thailand that they were *refusing* to perform funeral services of COVID-19 victims, the head of Thailand's Department of Medical Services *inappropriately* announced on March 25 that the disease was not contagious in bodies after death, to allow the funerals.

In light of this finding, forensic scientists should take several precautions while examining the remains of COVID-19 patients. For instance, they should wear protective gear, including a protective suit, gloves, goggles, a cap, and a mask.

'The disinfection procedure used in operation rooms might be applied in pathology / forensic units too,' they added.

WHO's view is different. Usually, pathogens that kill people don't survive long enough to spread to others after the person's death. Human remains only pose a substantial risk in a few special cases, such as deaths from cholera or haemorrhagic fevers, such as EBOLA.

Other illnesses that are contagious in human remains include tuberculosis, bloodborne viruses (such as hepatitis B and C and HIV), and gastrointestinal infections (including *E. coli*, hepatitis A, Salmonella infection, and typhoid fever), according to the WHO.

A lot of deaths have already occurred from COVID-19 in India, and Hindu dead bodies have been cremated. We still haven't heard of the virus erupting form a dead body of the affected patient, either because of touch or through the fumes or air surrounding the cremation grounds. Good for us.

**From Dogs (intestines)**

The novel coronavirus presumably originated in bats, but the pathogen may have then hopped into dogs before infecting humans, a new study suggests. According to Xuhua Xia [26], who scanned the genetic code of SARS-CoV-2 for a specific feature known as a CpG site, a sequence of genetic code in which the compound cytosine (C) is followed by the compound guanine (G).

The human immune system sees CpG sites as a red flag, signalling that an invasive virus is present. A human protein called Zinc-finger Antiviral Protein (ZAP) latches onto the CpG sites on the viral genetic code and recruits help to break down the pathogen.

Essentially, to survive and reproduce, a pathogen like COVID-19 needs the ability to evade the host's immune fighters, and in this case, it would mean getting rid of CpG sites smartly, that alerts ZAP proteins to the virus.

**From the Cats**

After closely analysing SARS-CoV-2's genetic lineage, the study's fresh authors say that SARS-CoV-2 combines the deadliness of the first SARS virus that emerged in China in 2003 with the

---

[26] Biology Professor **Xuhua Xia** of the University of Ottawa in Canada was born in Jiangxi, China, obtained his PhD in population biology from University of Western Ontario

contagiousness of HCoV-HKU1, a super contagious human coronavirus strain that hardly causes any symptoms at all.

Unsettlingly, SARS-CoV-2 is a mixture of the worst that both of these strains had to offer.

SARS-CoV-2, just like other coronaviruses, enters a new host's cells through its spike protein. The research team at Cornell discovered a new structural loop within SARS-CoV-2's spike protein that houses *four amino acids* that are quite distinct from any other known human coronavirus. Professor Whittaker believes that this structural loop and sequence of four amino acids are what make SARS-CoV-2 so contagious.

According to him a recent research did identify a *bat* in China carrying a coronavirus strain with a very similar structural loop to SARS-CoV-2, but the amino acid sequence wasn't the same. But their study confirmed that SARS-CoV-2 did originate in bats. Other studies have pointed to pangolins as being the animal originators of the virus, but the team at Cornell says there isn't enough evidence to support that theory.

The discovery of this new *amino acid sequence* suggests, that besides primates, *cats*, ferrets, and mink are the animals most at risk of carrying SARS-CoV-2. For the virus to infect a new organism, its spike protein must bind with a receptor located on a host cell's surface. Cats' receptors are very similar to humans. Feline SARS-CoV-2 infections, though, are very mild and rarely show themselves at all. Moreover, there is no evidence that cats can infect humans.

"We are keeping an open mind to see if similar things may happen in *cats* that already are now happening in humans," Whittaker concludes.

## Conspiracy Theories

Some content that appeared on YouTube, Facebook, WHO journals, Live science and Nature or in TED talks are good guesses of what different people are thinking of COVID-19. These posts are more or less of a permanent nature, meaning these will remain available for people to see.

The idea is not to promote or defend any controversy but to make people aware of all the innovative possibilities. Here are some thoughts, each one heavily contested by the group that promotes the controversy.

### It's only a virus – but deadly

We initially placed COVID-19 at par with the current viruses like hepatitis, HIV, dengue, brain fever, mad cow, and Spanish flu which have been killing the masses. But shouldn't we remember every new virus has peculiarities? And more significantly a new virus doesn't bring an antidote vaccine with it.

The closing of all the airlines of the world was not an insignificant event, it scared us like a bomb was about to be detonated. And in fact, it was.

The virus they say in animals is more potent or toxic than in human beings. Generally, they are transmitted from animals to animals or humans to humans, meaning separately. The problem occurs when cross-pollination or transmission takes place.

And this starts whenever a human eats an uncooked or partly cooked animal, and more dangerous when wild animals are involved. And maybe when an infected human is bitten by an animal, eg rabies virus.

The COVID-19 is assuming alarming reactions as now it has started showing in animals. The recent press report that a tiger in New York got it from the zookeeper/handler who was tested positive, raises our hair. A couple of days later two dogs and a cat were diagnosed positive in Hongkong. Whether a human carrier was involved in it or no, is not yet reported. I hope all this turns out to be a prank of April fool day.

Ultimately COVID-19 has all it needs to make it unfathomable and fatal.

Certainly, a manmade virus can be inflected by foreign genetic material by a process called transfection. [27] Transfection is the process of deliberately introducing naked or purified nucleic acids into eukaryotic cells. If the virus spike has been thus modified by the transfection process then it better be good enough to be attracted by the human ACE2 receptor. For successful cross-pollination, both the virus spike and the receptor must have a substantial affinity for each other.

Let me add a caveat to the above. Scientists have very recently discovered six entirely new coronaviruses lurking in bats in Myanmar. These viruses are in the same family as the SARS-COV virus that is currently spreading across the globe, but the researchers said the newbies aren't closely related genetically to SARS-CoV-2.

**Biochemical attack precedes the economic turnaround**

One persistent myth going around is that this virus was made by scientists and escaped from a lab in Wuhan, China, where the outbreak began.

The controversy that Corona is akin to an atom bomb exploding or a third world war being slipped in through a process where not even a single bullet was fired, is plausible only if this virus is manmade or

---

[27] Read **Transfection** in Wikipedia, it may indicate to you how externally a DNA can be modified

a laboratory construct. Or is it a mere accident being blown out and developed into a catastrophe.

*Is it true?*

Danielle DiMartino Booth CEO Quill was interviewed by W Patrick Bet-David, on 10-Apr-2020 [28] in which she called coronavirus an *Act of War*. Being a shrewd economist, she presented startling facts about the business deals signed between the US President and the Chinese Premier, wherein a clause was cleverly inserted by the Chinese, 'if there is an act of **God** or a **pandemic**, the Chinese will not be liable to honour their buying commitment from the USA'. Quite smart, isn't it.

This was at the beginning of Jan-2020, six weeks after the virus had surfaced in Wuhan. Why was its outbreak kept as a secret? Why the number of reported deaths was much less than the actuals? A toddler can tell you what happened subsequently in Germany, Italy, and Spain, why that toddler could not say it for China? Why were the travellers allowed to board the planes to Europe?

Substantial suspicion is gaining ground, as highlighted by WION (World is One News) an international news channel owned by Zee Media in India. It is somewhat astounding that a TV channel is permitted to broadcast several offending or felonious photos and clips in support of its opinion. WION asked why the Chinese director of WHO is not being questioned by anybody on the failure to activate precautions, as was done in the case of EBOLA.

Objections were made by the Chinese and the fact that this channel was selected to be banned for view in China, might support this theory.

Why were the funds withdrawn from WHO by the US government, at the time of this catastrophe when WHO is expected to continue to

---

[28] https://www.youtube.com/watch?v=jsV_YXq-1x4&feature=youtu.be   the YouTube interview

work? These questions are enough indications that something is not right between the US government and the Chinese counterparts.

*How it was executed*

The crux of this controversy is that the Chinese created this virus purposefully or hit upon it inadvertently. A Chinese doctor, who was arrested, removed from the hospital and who died with similar symptoms, has been quoted as the one who let the cat out of the bag.

As the gossip has it, they used their people, and some Americans who were working in Wuhan, to undertake international travels. Chinese presumably sacrificed a few locals in Wuhan as the bargain for millions of dying abroad. If the biochemical bomb was to explode and its fall out affected some smaller countries as well, but they didn't care.

It is rather strange that the Chinese shut their domestic flights but continued with international flights. Isn't it rather strange that Beijing and Shanghai, which are closer to Wuhan than Rome, New York, Madrid, or Venice, have reported almost negligible incidences of Convid-19? To make it look real, they published photos of the grounded Cathay Pacific fleet. Here too the lead was well administered, it was the first airline to go off the air.

*The culminating effects*

The WION channel goes ahead showing how the Chinese manipulated the economy by underplaying the stocks of foreign companies in China and let these companies' stock slip into alarmingly low rates. Subsequently, they acquired substantial shares of those companies and now hold the lion in them.

When the control of companies came in Chinese hands, they leapfrogged to commence the manufacture of protection kits for the virus, such as gloves, masks, and now ventilators. The current demand for these items from many countries is in millions. And slowly but surely, the Chinese are opening up their travels, bars, and pubs for their people, when the other countries are besieged and continuing with their sojourn of lockdowns.

The Indian government has taken steps to curb the full takeover of companies like Flipkart, HDFC financials, and PayTM by the Chinese.

Their first attempt where they failed miserably was the Great Leap Forward (1960) but now it seems the Chinese Government succeeded in fiddling with the international economy. They have probably won the Third World War without firing a single bullet.

*Partners in crime*

In the meantime, the FBI arrested a professor from Boston University who was in connection with Chinese university and research lab in Wuhan and was highly paid by China. A couple of scientists who either provided direct support to Wuhan technical university or were working in the US or were facilitating some contractors were also questioned. It was broadcast by the US government officially.

Somebody in the USA also planted a 3 trillion suite on the Chinese government. The reactions, don't they reflect suspicion in the minds of the US government.

[29] The state of Missouri has become the first US State to sue China, alleging that Beijing suppressed information, arrested whistle-blowers, and denied the contagious nature of coronavirus that led to the loss of lives and caused 'irreparable damage' to countries globally. Missouri's Attorney General Eric Schmitt filed the suit.

---

[29] **US State** files lawsuit against China on COVID-19 handling

**Read more at:**
https://economictimes.indiatimes.com/news/international/world-news/us-state-files-lawsuit-against-china-on-coronavirus-handling/articleshow/75286051.cms?utm_source=contentofinterest&utm_medium=text&utm_campaign=cppst

Senator Ben Sasse, a member of the Senate Select Committee on Intelligence welcomed the lawsuit.

It was reported on Live Science that, 'If it is a war, and I believe that's a proper metaphor, then we should fight it like a war. That means we should fight to win to vanquish the foe, not to let it persist and hassle us for an indefinite period.' Were they referring to China or COVID-19?

But they are amused by the fact that 'If someone were seeking to engineer a new coronavirus as a pathogen, he would have constructed it from the backbone of a virus known to cause illness. They wouldn't have chosen mutations that computer models suggest. Because the models don't tell us all about the mutations required.

But as it turns out, nature is smarter than the scientists, and the novel coronavirus found a way to mutate that was better - and completely different - from anything scientists could have created.

### Radiation hazard – a ghost or bogle

Chernobyl, Nagasaki, and Hiroshima were nightmarish calamities enough to send chills in our spines. So, why is it necessary to throw up this 5G controversy? Is it to scare us blue? Or is it for diverting our attention from COVID-19 to the radiation bogey? Or is it to underplay the 5G (5th Generation of mobile phones) spectrum?

The examples the promoters of this controversy quote are the proliferation of radars during the second world war and the recent spread of 5G communication satellites.

For a long time, scientists were cautioning people on the dangers of radiation hazards on human bodies. We were quick to agree to carry different colour bands when visiting nuclear reactors and the like, depending on the intensity of radiation expected in a designated enclosure.

The current use of mobile phones could also be the cause of radiation. But that cannot be extended too far.

Some confusion between *ionizing* and *non-ionizing* radiation exists because the term radiation is used for both. All light is radiation because it is simply energy moving [30] through space.

The ionizing radiation is dangerous because it can break chemical bonds. Ionizing radiation is the reason we wear sunscreen outside because short-wavelength ultraviolet light from the sky has enough energy to knock electrons from their atoms, damaging skin cells and DNA.

5G uses non-ionizing millimetre waves. They're called millimetre because their wavelengths vary between 1 and 10 millimetres, whereas radio waves are on the order of centimetres. The millimetre waves use frequencies from 30 to 300 gigahertz, which are 10 to 100 times higher than the radio waves used today for 4G and Wi-Fi networks.

Millimetre waves have longer wavelengths and don't have enough energy to damage the human cells directly. The only established hazard of non-ionizing radiation is excessive heating.

At high exposure levels, radio frequency (RF) energy can indeed be hazardous, producing burns or other thermal damage, but these exposures typically incur only in occupational settings near high-powered radio frequency transmitters, or sometimes when medical procedures going awry.

But there's one problem though, and a blessing as well, Millimetre waves are easily absorbed by foliage and buildings and will require many closely spaced base stations, called small cells, to establish seamless connections. Fortunately, these stations are much smaller and require less power than traditional cell towers and can be

_____

[30] Reference Live science - 5G

placed atop buildings and light poles. But this is a rather welcome concept, a blessing in disguise.

The miniaturization of base stations enables a technological breakthrough for 5G: Massive MIMO. MIMO stands for multiple-input multiple-output and refers to a configuration that takes advantage of the smaller antennas needed for millimetre waves by dramatically increasing the number of antenna ports in each base station. (picture shows MIMO).

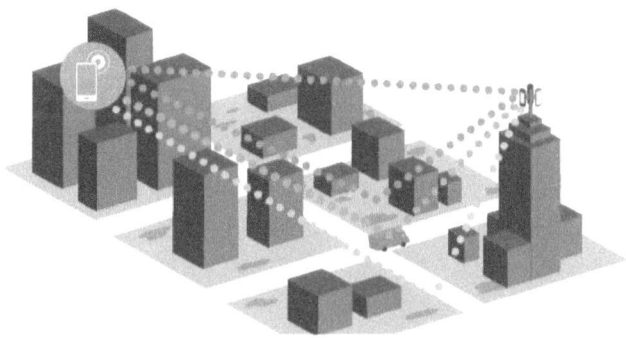

Millimetre-wave and massive MIMO are the two biggest technologies 5G will use to deliver the higher data rates and lower latency we expect to see.

What could be the result of these concerns over the 5G's use of the higher energy millimetre-wave radiation, will be clear only after its complete deployment. Although some detrimental indications about the microwave ovens at homes are emerging.

For the present, we can ignore this controversy.

**Temperature belt or ribbon across the globe**

Some isolated but concerted studies have been made (not by WHO or CDC) that the spread of this virus is confined to a ribbon of stretch about 200 KM wide that spans the whole globe. Temperature, humidity, or microwave data table on different countries or parts thereof, support this claim.

It is said by the believers of this theory that the coronavirus melts in higher temperature, gets frozen, and stays dormant or becomes noticeable in human bodies. This argument is being advanced as the summer of 2020 comes around and people are counting hopes in all countries such as the USA, India, etc.

The virus is susceptible to ultraviolet (UV) rays, and these will be easily approaching due to clear skies and bright sun. High relative humidity has been found to affect virus survival. Externally it may help as well. But dry ambient air reduces protective mucous production in the respiratory passages. This will certainly be detrimental inside the body.

Though the possibilities discussed above may indicate that Indians may escape COVID-19's fury, the actual impact will be clear in the next few weeks.

### Skin Colour

Look at the WHO data tables for COVID-19. Most of the population affected is white and the least is in Africa and dark skin countries. In contrast, EBOLA affected only dark skin people.

Dark skin people are accustomed to medicines on malaria, dengue, and influenzas, etc. They seem to be receptive to chloroquine etc. They have a presence of the built-up antibodies inside their bodies. Until today the number of cases detected in these countries is a small percentage of the population compared to other countries.

Are people with white skin more prone statistically? Although the US President made a contradictory statement just a few days back, he indicated more incidences in dark skin people.

## Commentary on conspiracy theories

The world has seen innumerable deaths due to pandemics and we take it for granted as if it is the will of God, but imagine the similar number of people dying due to an atomic or hydrogen bomb, or by a terrorist using his grenades, will it not be a ghastly act.

Globally, many people are affected and a subset has died. Yet It is important to note that many people to the order of 896,000 have recovered as well, which is something to be reservedly proud of. Compared to the other disasters described in this book, and keeping our present-day population numbers in mind, the death numbers, 211,350, are not so alarming so far.

Corona is a great leveller, an equalizer of beliefs, and that includes religions, political paradigms, the colour of the skin, and regions of the globe.

Yet, if these controversies are driven from legitimate concerns it's ok, but if these are subtle ways to promote any individual or an agency or a journal, it is not excusable.

*But my ultimate question remains. Why does the coronavirus spread by hand touch, as is being said? Why a hard surface is a cosy bed for this virus whereas a fluid medium should have been more conducive. How can washing hands contribute to its disappearance or its death? Would it be simple if we state that corona is in the air so, don't breathe?*

You will catch it in the flying jets or moving trains
or in the deep seas if you are scuba diving
**along with a carrier**.

THE
# CORONA
SYNDROME

A chance for introspection

MIKE RANA

# 9. Tackling COVID-19 – All that we could do

$A$ll of our economic, medical and academic resources are currently occupied by the COVID-19 disaster. Yet dialysis patients still need care, cardiovascular disease remains a prevalent killer, suicide rates are still rising, tuberculosis and cancer haven't gone away. The lower-and middle-class citizens are still square pegs falling off round holes. All of these should not be ignored. Who knows somebody may be on the edge, the line between life and death?

By the way, don't you think it is rather odd that all Outpatients Departments OPD) in hospitals are shut for the last few months, and corona patients have bombarded them? What does it mean? How do the people who generally line up for meeting doctors daily have out-of-the-blue become healthy? Is it not necessary to look deep into the flow of money from the patient to insurance companies to the hospitals and finally getting added to the doctor's commission?

The United States did begin to prepare for the storm before the calm. Social distancing - the word - was introduced and as copy cats, we started talking about it in our country as well. But the US faltered a bit. Even though it was introduced and people said it was too late, it wasn't even implemented strictly. People's attitudes and relaxed callousness of the police didn't help either.

So far, experts are not synchronised on the approach that we should be taking. One needs to find a common denominator for the medications. But still …

## Early Detection – the key to preventing an outbreak

*A stitch in time saves nine.* I am a die-hard fan of this proverb and have used it a lot to annoy my wife. No time could be more appropriate for its use than now. Fundamentals dictate the following for early detection.

**The Test Kit**

Firstly, and the most ground-breaking way is simply a dipstick or litmus paper type test. When available it will be easily administrable, cheap, and quickly available (in all chemist shops, Amazon, and the hospitals). Won't we prefer to be served with the likes of diabetes or urine testing probes that enable full uncompromised privacy of people? Developing a testing probe, therefore, takes *top* priority.

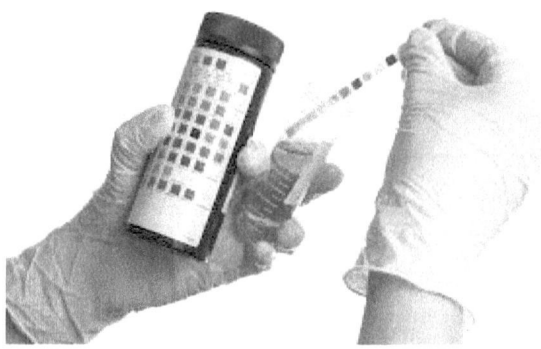

On COVID-19 testing kits, the Indian CSIR  chief said on 25-Apr-2020, that in 48 hours two new rapid testing kits have been developed - one by the institute and the other by Sree Chitra Tirunal Institute.

While the kit developed by SCTIMST is an RT lamp, the CSIR-Institute of Genomics and Integrative Biology has developed a paper-based kit. Both the kits produce accurate results in 30 minutes, Mr Mande claimed.

**Malfunctioning of Chinese test kits**

It has been reported by Chief Ministers of different Indian states (led by the opposing parties) Rajasthan, Punjab, West Bengal that the test kits from China failed to achieve the desired results. For example, they failed to confirm not more than 10 % from the confirmed lot of COVID-19 cases.

The Haryana government cancelled the Test Kits order from China and is now waiting for the refund money. They placed a fresh order on a South Korean company based out of Manesar, an industrial town near Delhi.

[31] Dr Trehan of Medanta hospital expressed serious concern on national television last fortnight about the failure of Chinese test kits. He expressed faith in the test kits from Rhodes Pharmaceuticals USA.

## Test centres

The second more feasible option is to reduce the *cost* and *time* for a traditional test in a hospital. But we should remember that in India, particularly in small cities, the hospitals are ill-equipped where even taking a blood sample or a swab from the throat is a challenge.

And then how many test centres can we marshal quickly. For an aggressive campaign like South Korea's one, we probably need 20,000 test centres to cover the whole of the Indian subcontinent, so was stated by senior South Korean diplomat Yoo Chang-Ho.

But what our concern is, what if we fail to transfer the samples to tests centres quickly. The solution to this was initiated by the government by upgrading the facilities at all hospitals in India.

At the time of penning down this book, we had 235 government and 84 private labs testing for COVID-19. They have so far tested over four lakh samples. More than 600 hospitals have been earmarked to handle COVID-19 patients.

One cannot comment if these numbers are enough or less for the kind of population we have in India. These numbers should be computed per cent basis, to get a better perspective of things.

---

[31] **Dr Naresh Trehan**, Chairman & Managing Director, Medanta, Gurgaon

A third option is arranging Drive-Thru mobile test centres for which vehicles are required, besides.

Yoo Chang-Ho also said it was important for the government to stay transparent while seeking the citizens' support for a mass testing program.

**Masks and gloves**

Masks, gloves, and other PPEs have been under debate right from the onset of COVID-19, yes, no, partly, specifically designed, re-usability, etc. Even the quality of PPE deployed has been in question.

Surgical masks can help to prevent the infected people from spreading the virus further by blocking any respiratory droplets that could be expelled from their mouths.

However, as the knowledge trickled in, it became apparent that the asymptomatic [32] people could be spreading the virus. Up to 25% of people with COVID-19 may fall into this category. Moreover, a new small study found that COVID-19 may be most infectious when symptoms are mildest, meaning that people may be spreading the virus before realizing they have it.

'In light of new shreds of evidence, CDC recommends wearing *cloth* face coverings in public settings where other social distancing measures may be difficult to maintain.' Under the new recommendation, healthy individuals are advised to wear mouth and nose coverings - including homemade masks, scarves, or bandanas - when they go to a public area, such as the grocery store or a pharmacy.

---

[32] A disease is considered **asymptomatic** if a patient is a carrier for a disease or infection but experiences **no** symptoms. A condition might be asymptomatic if it fails to show the noticeable symptoms with which it is usually associated.

As before, the CDC doesn't recommend the public to wear *N95*  *respirators*, which presumably filter out 95% of particles in the air. These masks are in short supply and should be reserved for health care workers who are exposed to the virus on a 24x7 basis. Nor should the public wear surgical masks, which are also needed by healthcare workers.

If you wear a homemade cloth mask, scarf or bandana, remember to wash your hands before you put it on. Also, you should wash cloth masks after each use and to always put the same side against your face, so you're not placing the outside, or the 'contaminated side' against your mouth and nose, according to Anna Davies, a researcher at the University of Cambridge.

An extract from friends Ira Khanna and Tasneem Lokhandwala from Facebook contacts.

*"I wear a mask in public, not for me, but YOU. I want you to know that I am educated enough to know that I could be asymptomatic and still give you the virus.*

*No, I don't 'live in fear' of the virus, I just want to be a part of the solution, not the problem. I don't feel like the 'government is controlling me', I feel like I'm a contributing adult to society and I want to teach children the same.*

*Wearing a mask doesn't make me weak, scared, stupid, ugly or even 'controlled', it makes me considerate.*

*When you think about how you look, or how uncomfortable it is or what others think of you, just imagine someone close to you...A Father, a Mother, Grandparent Aunt or Uncle choking on a respirator or a ventilator. Then ask yourself if you could have sucked it up a little for them."*

## Personal Protective Equipment (PPE)

Substantial effort is required for procuring the Personal Protective Equipment (PPE). Hitherto it was imported but more significantly the local laboratories and industries have been roped in to manufacture these within India, using indigenous textiles. Good effort.

## Social distance – how much is that distance

Social distancing is a brave commitment that in a time of crisis our nobility and resolve considers the health of those around us. Those we may not even know, the ones who appear in our lives just as headlights in the other lane of the highway or even the elderly people that exit the train a few stops before us. It's about physical distance.

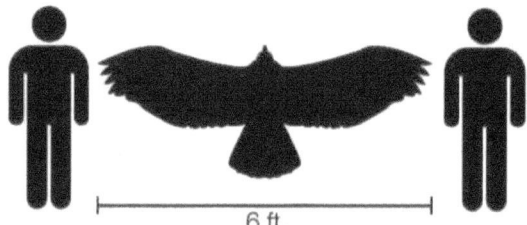

This idea, that large droplets of virus-laden mucus are the primary mode of transmission guides the CDC's advice to maintain at least a 6-foot distance between you and other people.

The thinking is that gravity causes those large droplets (which are bigger than about .0002 inches, or 5 microns, in size) to fall to the ground within a distance of 6 feet from the infected person.

But that 6-foot guideline is more of a ballpark estimate than a hard and fast rule, said Joshua Santarpia. That's because even large respiratory droplets can travel fairly far if the airflow conditions are right.

'There isn't anything magic about standing 6 feet away from someone that you are interacting with directly. If you stand to talk to someone who is infected with the virus, whether it's 3 feet or 6 feet, there is going to be some risk of infection,' Santarpia told Live Science in an email.

Also, some experts believe the 6-foot rule is based on outdated information.

'At this point, I don't think anyone can take a group of people with COVID-19, and say how each person has become infected, and then say that xx% got infected with droplets and yy% got infected via touching surfaces,' Dr Jeffrey N Martin [33], told Live Science.

'6 feet are probably not safe enough. The 3-6-foot rule is based on a few studies from the 1930s and 1940s, which have since been shown to be wrong - droplets can travel farther than 6 feet,' said Raina Macintyre. 'Yet hospital infection control experts continue to believe this rule. It's like the flat Earth theory - anyone who tries to discuss the actual evidence is shouted down by a chorus of believers.'

Another complicating factor is that at least 25% of the people who are transmitting the virus may be asymptomatic at the time, said Dr Robert Redfield. [34]

---

[33] **Dr Jeffrey N Martin**, a professor in the Department of Epidemiology and Biostatistics at the University of California, San Francisco

[34] Dr **Robert Redfield**, director of the Centres for Disease Control and Prevention

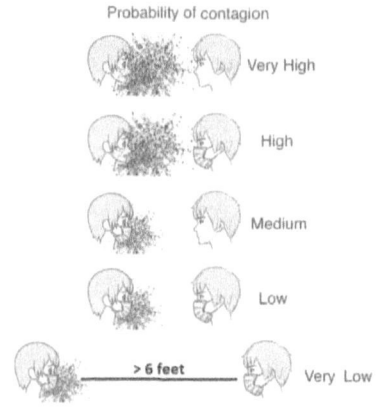

Probability of contagion

Very High

High

Medium

Low

> 6 feet

Very Low

## Is takeout food safe?

So far, there's no evidence that the virus is transmitted via food. The virus will not live long in food per se, and while it's possible that food packaging from groceries or takeout could contain small concentrations of virus particles, it is easy to mitigate this risk by washing your hands after handling groceries or takeout, Ben Chapman says. [35]

## Take good care of your unaffected family members

The fact that so many seemingly innocuous activities can transmit the virus can be scary. And it can be even scarier not knowing the actual risks associated with each transmission route - without that information, how can we take the right steps to protect ourselves?

'What is true is that persons who have a member of their household infected with the virus have a higher probability of getting infected than people who don't have such close exposure. This tells us a lot. This tells us that *close contact* is the most important factor,' Martin said.

Briefly passing a person on the street, at a distance of 6 feet, is likely to pose a risk but it is low, chatting at a distance of 6 feet with that same person for a few hours will be a higher risk.

## Categorize and segregate people

According to me, the population should be divided into five groups. This could be taking place at the State level.

---

[35] **Ben Chapman** a professor and food safety specialist at North Carolina State University.

- **Category 0** - Those who are **not** known to have been exposed or infected. Nothing except quarantining at home may be required. More if their residences are located in an *epicentre* or *hotspot* district.

  But we need to be extremely careful with these (majority of) people, because they might have infected others (probably within their families) and will themselves get affected later. We call these people as asymptomatic carriers. The irony is they might have infected others in their surroundings.

- **Category 1** - Those **infected** with COVID-19 (*primary, secondary* and *tertiary*) - These should be hospitalized - if they are very sick - or placed in *'infirmaries'* (such as a converted convention centre) if they have mild to moderate disease.
- **Category 2** - Those **presumed** to be infected - based on symptoms but who initially test negative. These could be placed in infirmaries and re-tested after one week.
- **Category 3** - Those **exposed** to someone – but don't yet show symptoms but they report problems, maybe quarantined in hotels or homes for two weeks. An easy case to handle.
- **Category 4** - Those who recover from COVID-19. These are, in theory, immune, and may be able to go back to work. This category, which would require the use of antibody-based tests to identify, could be a game-changer in restarting parts of the economy more quickly and safely. [36]

We probably need to develop a flow chart that could be broadcast to everyone in the public domain eg on social networks like Facebook, Instagram, WhatsApp, or Twitter.

---

[36] According to the reports in Guardian, researchers in **Germany** have already started a large study to find out how many people in the country are immune to COVID-19, which could allow officials to issue 'immunity passes' to allow people to return to work.

As soon as you suspect to be in *category 1 to 4*, start maintaining a diary of symptoms. The diary should record data on a 3-hour time slot, and this should be at fixed times

- Keep yourself aware of the reliable hospitals nearby
- If you change over to *categories 2, or 3*, Decide what date and time you need to get out of your home quarantine and go to the hospital instead.  And who will accompany you and in which transport
- Carry all the documents that you need for example your Aadhar card and your relevant medical history, and leave your contact number for emergency calls
- If the hospital doesn't have the requisite facilities for the test, they will have to send/carry the swab or blood sample to a test lab. To ensure a quick response, therefore, reach the hospital in the morning times so the samples can be despatched the same day, or else go to the lab directly
- They will probably need the medium for keeping the sample, in transit, or frig. [37]
- The result will be received after 3 days, either from the test lab or from the hospital. With time this time may be reduced to a few hours. During this time the patient will have to be treated, housed, or quarantined. One must be prepared for all eventualities.
- If the patient feels more discomfort after returning home, he should consider he has gone back to Category **2** and he should report back to the hospital immediately

### Do's

The *cost* of confronting the coronavirus pandemic is millions of jobs and trillions in wealth lost to *save* potentially millions of lives. The

---

[37] **Medium for Transit**: When a swab or blood sample is taken the medium where it is enclosed is very important it should not be thin air and the liquid or fluid is specially manufactured and stored and shipped. This is a caution for the logistics staff connected the testing.

world has always managed to find ways to improve itself in the midst of - and sometimes because of - its most difficult challenges.

A Chinese expert assures everybody that inhalation of steam from hot water kills coronavirus 100 per cent. Even if the virus enters the nose, throat, or lungs. Corona Virus cannot stand the hot water steam. Is it true, as some other claims by the Chinese create doubts in our mind? Are we being proposed to be used as guinea pigs?

And there is a growing belief the heat kills the virus. All over the world, this is the trend now.

Maybe the following bullets during the Lockdown or exit afterwards will help

- Don't think that Corona will never come to you
- Stay at home, this is all that you require
- Follow the advice on personal preventative measures (respiratory etiquette, hand washing)
- Quarantine yourself or your family members who return from abroad, at home and stay away from them
- If an outsider comes to your home, make sure that you don't go anyway near him
- Ensure people with symptoms or with family members with symptoms go ahead with self-quarantine for 14 days
- Check the body temperature of employees daily so that employees with fever can be prevented from coming to work
- Provide alcohol-based gel or washing hands stations at easily accessible points
- Remember medical services are already filled to the brim, don't overexert them
- For the medical workers, like doctors and nurses and all the others, when they return after a long day at the hospitals a few extra precautions are necessary
  - Firstly, the shoes must remain outside the house
  - Wash hands thoroughly on all sides

- o Remove clothes and put them away for a normal wash to be dried in the sun outside
- o There is no need to have a shower with Dettol or other antiseptics, just a plain shower with adequately applied soap is enough
- Sleep and give rest to your mind, but don't laze around
- Exercise a lot keep your body trim and active. Don't gain unnecessary weight
- Get closer to children and establish bonding. If network games are played select the educative ones like brain teasers etc
- Drive your hobbies, like writing (as this book was written in 6 weeks) or/and reading books
- Improve your professional skills as well as hobby-oriented skills, and there is a huge list of these
- Don't get angry on trifles

Violation may involve penalties or fine but **no one** will be able to get your life back

## Don'ts

- Spread rumours or post fake news on social network
- Refrain from spreading explicit videos on something that impresses only you
- Stay outdoors or go to public places like parks and streets
- Go to the unapproved shops or offices
- Overeat at home
- Watch too much of Internet thereby clogging the network
- Panic buying and stockpiling food grains, it does affect the poor, the elderly, the less abled, and those with other ailments and circumstances.

## Keep them away - Social Network Distancing (SND)

Did you ever think that social distancing and social *network* distancing (SND) are two different terms?

A subject hitherto ignored. Social network distancing involves reducing the network traffic and leaving adequate free space or bandwidth for Work-From-Home (WFH) operations.

It can be done by reducing your posts, selfies, virtual friends, your business ads which probably don't get you the expected response, and above all the unnecessary repetition of video clips particularly those already available on YouTube, etc.

People get offended or outrightly angry seeing others posting un-confirmed and scandalizing reports on the internet, remember that you might be doing exactly the same. Stay away from being a spam writer.

Of course, there's a blur between personal and work use. Currently, people are starting their day maybe earlier, so the workday is longer. When you are at home longer you are doing more activities on your phone or computer. The load on the network increases as you start doing video conferencing. It hogs the bandwidth.

No wonder network operators and specifically the Internet Service Providers (ISP) have confirmed the surge or have started complaining. They shifted to the *overdrive* gear, yet it's a testament to them that the networks are holding up rather well.

As of a few weeks back, AT&T reported, 'Wireless voice minutes of use was up 39% compared to an average Monday. Wi-Fi Calling minutes of use were 78% higher than an average Monday. Consumer home voice calling minutes of use was up 45% from an average Monday.' The company said its core network traffic, including 'business, home broadband, and wireless usage,' was up 27% on Mondays compared with the same day last month.

VPN traffic was up 52% over a typical day on Verizon's network.

## Release after recovery

### CDC Guidelines for release after recovery

The duration of time individuals remain contagious might be related to case severity. The WHO states, 'for people with mild disease, recovery time is about two weeks, while people with severe or critical disease recover within three to six weeks.'

These guidelines were issued by the CDC recently about when a patient can be released from isolation.

- The patient doesn't have a fever anymore even without the use of fever-reducing medications.
- The patient is no longer showing any symptoms, including any cough.
- Patient obtaining negative tests on at least two consecutive respiratory specimens collected at least 24 hours apart.
- Another question that could take years to answer is whether the SARS-CoV-2 virus that causes COVID-19 may lie dormant in the body for years and spring back later in a different form or attacking different organs.

### Published in Los Angeles Times

Scientists in China examined the blood test results of 34 COVID-19 patients throughout their hospitalization. In those who survived mild and severe disease alike, the researchers found that many of the biological measures had *failed to return to normal*. Eg impaired liver function. When lungs do a poor job of delivering oxygen to the body, the heart can come under severe stress and may emerge weaker. [38]

---

[38] https://www.latimes.com/science/story/2020-04-10/coronavirus-infection-can-do-lasting-damage-to-the-heart-liver? amp=true   Los Angeles Times

'COVID-19 is not just a respiratory disorder', said Dr Harlan Krumholtz, [39] a cardiologist at Yale University. 'It can affect the heart, the liver, the kidneys, the brain, the endocrine system, and the blood system.'

## Quarantining

Even if somehow, we can detect the presence of coronavirus in people adequate local doctors, medical or paramedical support isn't available for a substantial portion of the Indian population. But that's not something for which we should be embarrassed about. The story is the same in all countries and they are saying, 'Can we really test all these cases? Can we isolate all the confirmed cases?'

Facilities are therefore required to *quarantine* the infected person (some hotels and guest houses are being requisitioned for this function). This is in addition to some five-star facilities made available to the medical staff.

One innovative solution used in India was the conversion of railway beds in the compartments to hospital beds, not the best solution but better than putting people on the floors. I am not aware of this facility was deployed at all, though.

This is the situation today and this will be the situation forever, and in all countries for every *new future virus* detected. In India, when COVID-19 struck us, we kind of ignored it, but then lately we did better than most of the early contaminated countries.

Isolation and quarantine are public health practices used to protect the public by preventing exposure to people who have or may have a contagious disease.

- *Isolation* separates sick people with a contagious disease from people who are not sick.

---

[39] harlan.krumholz@yale.edu

- *Quarantine* separates and restricts the movement of people who were *exposed* to a contagious disease to monitor if they become sick. These people may have been exposed to a disease and do not know it, or they may have the disease but do not show symptoms.

## Medical options to be tried

Scientists around the world are running a race to find ways out of the current coronavirus pandemic. Some are working to develop new drugs and vaccines, while others are looking to see whether therapies, we already have may help against COVID-19.

### Ventilators – are they deadly

Nearly 9 in 10 COVID-19 patients who are put on a ventilator die, New York hospital data suggests. But age makes a difference.

Around 76% of ventilated patients between the ages of 18 and 65 died, and 97%, of ventilated patients over the age of 65 died, according to the report.

The bleak statistics don't imply that the ventilators caused harm, said senior author Karina Davidson. [40] Rather, 'patients who are put on ventilators typically have more severe disease,' and are therefore likelier to die, Davidson told Live Science in an email. 'Mechanical ventilators are not dangerous, and in many cases, are lifesaving machines.'

### Possibility of BCG vaccine

In the latter category, researchers have dusted off one intriguing compound in our collective medicine cabinet - a century-old vaccine to fight tuberculosis, a bacterial disease that affects the lungs. A couple of early analyses, which have yet to be peer-reviewed, have found that countries that have used the time-tested vaccine, Bacillus

---

[40] **Karina Davidson**, senior vice president and professor at the Feinstein Institutes for Medical Research at Northwell Health.

Calmette–Guerin (BCG), seem to have been hit less severely, in terms of both number and severity, by coronavirus.

According to [41] Dr Randip Guleria, the Indian population is given BCG protection almost invariably in their youth. Going by the renewed initiative, some patients could be administered BCG again to nullify the fear that BCG has lost its effect. It could at the most be a sample test.

**Probability of plasma therapy**

Recently it was opined and tested in the USA that if a patient who has recovered from the corona disease is agreeable to donate his *blood* then there is a possibility of using his plasma, as a cure. This has been found in India as well. This technique was also used in other previous virus incidents.

For this, the blood plasma should be extracted, tested for the absence of other viruses like HIV or Hepatitis, and injected into a serious patient, provided the plasma is infection-free. It is expected that such injected dosage would generate or multiply the antibodies required to fight the virus.

This is something thing that Bill Gates [42] had mentioned long back in his TED talk. He had proposed that pending the creation of a new

---

[41] Prof **Dr Randip Guleria** director AIIMS Delhi

[42] As is said Bill has invested 200 M dollars into conglomerates like Schering, Wyeth Pharmaceuticals, Gilead, Boots, Walgreens, Johnson and Johnson, Pfizer, Bristol-Meyers-Squibb, GlaxoSmithKline, Merck, Lilly etc. The Coalition of Epidemic Prepared Innovation (CEPI) sponsors Moderna for finances and coordination of development of vaccines.

Moderna uses the facilities of Innovia, Wistar institute, VGXI GeneOne Life science, and Twist bioscience. One thing is common in all of these. All financed by **Bill&Mellinda** foundation for a vaccine against SARS-Cov-2 virus.

They are into the business of creating vaccines on a commercial basis to make immense profits. For example, The Pillbright Institute applied for a patent for coronavirus vaccine on 20-Nov-2018.

vaccine, an alternate procedure could be used. The blood from affected patients could be used for treating future patients.

In my opinion plasma therapy is the closest to a vaccine, except the container for the medicine is different. The vaccine comes in a capsule and plasma in tubes.

**Probability of a vaccine**

Notwithstanding the aforementioned possibilities, there's been a lot of talk in recent weeks about a new coronavirus vaccine. But most of that chatter is nothing more than wishful thinking and, at best, projections or estimates.

While at it, we should be aware that many people and companies including the famous philanthropist Bill Gates, well known for his vaccination program, are sponsoring many programs on vaccine creation. And we should be expecting a COVID-19 vaccine in a few months.

*PittCoVacc*

In a bit of concrete good news, researchers from the University of Pittsburgh's School of Medicine announced the development of a new SARS-CoV-2 (PittCoVacc) vaccine that has shown promise in mice trials.

Within two weeks of receiving the vaccine, test mice developed a *surge* of protective antibodies against the COVID-19. Those mice haven't been tracked for a long period but researchers estimate that the mice should remain immune to SARS-CoV-2 for as long as a year. The vaccine, administered via a small finger-tip sized patch, shows promise.

The delivery method for PittCoVacc, on the other hand, is rather unique. Referred to as a microneedle array, the delivery patch contains 400 tiny needles that deliver spike protein pieces into the skin. While the idea of 400 needles is enough to make even the most fearless among us squirm, the patch is quite painless. The needles are very tiny and made completely from sugar. They dissolve into the skin and release the protein pieces. Researchers compare administering the vaccine to putting on a band-aid.

The PittCoVacc vaccine also retained its potency after being sterilized with gamma radiation. While that may sound like a process straight out of *Incredible Hulk* comics at first, gamma

sterilization is a key-process vaccine must go through before moving onto human trials.

*Chinese monkey virus-based vaccine*

Sinovac Biotech, a privately held Beijing based company is an experienced vaccine maker - it has marketed inactivated viral vaccines for hand, foot, and mouth disease; hepatitis A and B; and H5N1 influenza or bird flu.

For the first time, one of the many COVID-19 vaccines in development has protected an animal, *rhesus macaques,* from infection by the new coronavirus. The vaccine, an old-fashioned formulation consisting of a chemically inactivated version of the virus, produced no obvious side effects in the monkeys, and human trials began on 16 April.

This company gave two different doses of their COVID-19 vaccine to a total of eight rhesus macaque monkeys. Three weeks later, the group introduced SARS-CoV-2, the virus that causes COVID-19, into the monkeys' lungs through tubes down their tracheas, and none developed a full-blown infection. They gave differing dosages of the virus as well as a vaccine to monkeys, high, medium, and low to corresponding numbers of monkeys.

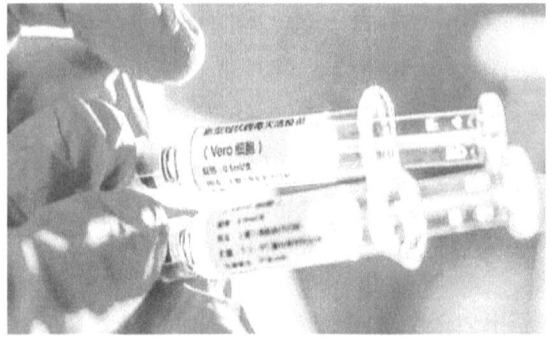 The monkeys given the highest dose of vaccine had the best response. Seven days after the animals received the virus, researchers could not detect it in the pharynx or lungs of any of them. Some of the lower dosed animals had a 'viral blip' but also appeared to have controlled the infection.

In contrast, four control animals developed high levels of viral RNA in several body parts and severe pneumonia. The results 'give us a lot of confidence' that the vaccine will work in humans, says Meng Weining, Sinovac's senior director for overseas regulatory affairs.

Knowing the history of Covid-19, Douglas Reed [43] says the number of animals was too small to yield statistically significant results. It may have caused changes that make it less reflective of the ones that infect humans.

The company hopes to start phase II studies by mid-May that have the same design but enrol more than 1000 people, with results due by the end of June.

---

[43] **Douglas Reed** of the University of Pittsburgh, who is developing and testing COVID-19 vaccines in monkey studies

*CSIR India's anti-leprosy vaccine Mw For COVID-19 [44]*

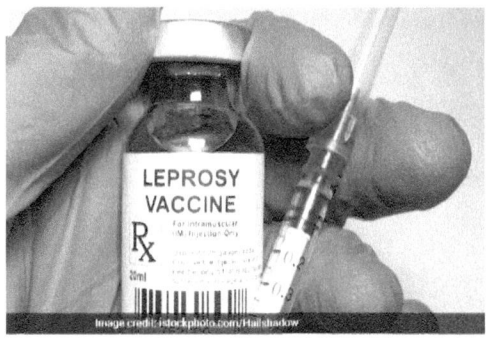

Dr Shekhar C Mande [45] on 19-Apr-2020 said that the institute is currently testing Mycobacterium w (Mw), a vaccine for leprosy, to treat COVID-19. 'The CSIR is concentrating on immune-boosting vaccines and hence Mw is being tested. He added, 'The Mw is usually used as a vaccine for leprosy. It is a micro bacterium species and we will be using a strain of it for an immune-boosting vaccine.'

CSIR has tied up with several hospitals and they are in the process of recruiting patients for clinical trials. As usual, Indians are innovative and creative, they were not going after the *vaccine* for the virus they tried *bacterial* way.

## Enabler application for 'contact tracking'

Another important issue is that, even if we get the number of infections down so that containment is possible, we will need an 'army of efficient, effective public health workers' to perform **contact tracing**. This involves tracking down contacts of patients, testing, and isolating them to stop this virus from *re-spreading*.

---

[44] **Leprosy**, also known as Hansen's disease, is a chronic infectious disease caused by Mycobacterium leprae. The disease mainly affects the skin, the peripheral nerves, mucosal surfaces of the upper respiratory tract and the eyes.

[45] Council of Scientific and Industrial Research (CSIR) Director-General **Dr Shekhar C Mande**

In Feb-2020, China launched a mobile application to deal with the disease outbreak. Users were asked to enter their name and ID number. The app could detect 'close contact' using surveillance data and therefore a potential risk of infection. Every user can also check the status of *three* other users. If a potential risk is detected, the app not only recommends self-quarantine, it also alerts local health officials.

Big Data analytics on cell phone data, facial recognition technology, mobile phone tracking, and artificial intelligence are used to track infected people and people whom they contacted in South Korea, Taiwan, and Singapore.

In Mar-2020, the Israeli government enabled security agencies to track mobile phone data of people supposed to have coronavirus. The measure was taken to enforce quarantine and protect those who may come into contact with infected citizens.

Also, in Mar-2020, many countries implemented similar techniques for contact tracking. 'The experience of COVID-19 in the city of Shenzhen may demonstrate the huge scale of testing and contact tracing that's needed to reduce the virus spreading,' explained Dr Ting Ma from the Harbin Institute of Technology at Shenzhen, China, in a press release.

*Aarogya Setu* is a recent application sponsored by the Health Ministry in India. It is being promoted by PM Narendra Modi and heavily advertised as well. It's no doubt a great effort but it is surely a great mission that it wants to achieve.

On using it personally I fund that it works on the data that you personally feed into it. It doesn't do any visible extrapolations on its own, except it gives you a final judgement whether you are safe or not.

Others users too reflected a similar sense of dissatisfaction, few of which are enlisted below.

- This App is draining my phone battery very fast. I don't think it is working properly because the place where I live, 3 corona patients have been found nearby, nearly 300 meters away. This app is showing different results on different phones like 'You are safe' on my phone and 'Low risk of infection' on my wife's phone. (*She must have visited outside*)
- One major drawback of this application is that there is no regular update of COVID-19 positives in India and its respective states. (*such updates are available elsewhere on the net*)
- Whenever I try to open the app it takes me to register window and then forward
- The app while taking the self-test asks a question 'Have you travelled abroad in the past 14 days?' With no international travel possible, since 23-Mar, the question is absurd. (*yes, but this will apply in the future*)

The application is released with half-baked tests or calibration. And this is why the developers are accepting the suggestions that it has real flaws. But that's ok since many applications have teething troubles.

But here are some suggestions, if someone is reading this book.

- Because of my earlier profession that of a software engineer, I can deduce that the software has been designed *without* specific approval of its *requirement specification* from the public users or doctors
- It should be an application like Facebook or Twitter but should be designed to work on Amazon Web Services (AWS) platform, which is the current trend on larger systems, and Aarogra Setu will in no way be a medium-sized application. This is the ultimate as on date. And without this, we would only be on a wild goose chase.
- Having taken care of databases, SQL or noSQL cloud databases, face recognition, or the encryption and security

services that AWS will take care off, all we need to focus on our application, which in our case is Aarogya Setu.

- And hopefully, our application will be based on Artificial Intelligence, and not on straight forward scanning of databases of infected people viz-a-viz the mobile phone user. It should use heuristic drilling techniques and cull before giving out the final message on the mobile phone
- Google can also give you forecast locations based on your history of movements
- Google has excellent face photo scanning features as well. One can trace out an infinitesimal face icon from a photograph of tens of people, quite reasonably well.
- Google can track your location while on move, with or **without your sim** inserted, and guides you to your destination

Sunder Pichai, Bill Gates and Melinda Gates, Mark Zuckerberg, Satya Nadella, are you listening. Amazon and Microsoft servers, will you please extend your benevolence towards this application, it could find use elsewhere in the world. Social distancing is the name of the game.

Indian Companies like TCS, Infosys, Wipro, and Mahindra don't lose this golden opportunity to dig down and extract your required data. Mukesh Ambani or Gautam Adani please come on board.

For more info: mikerana@gmail.com

## Myths – prevalent all over the globe

While all efforts are being made to tackle the coronavirus, here are a few myths that emerged and are floating about.

- Face masks can protect you from the virus; *not by itself*
- You're way less likely to get this than the flu; *not necessarily*
- The virus is just a mutated form of the common cold; *no, it's not*

- The virus was probably made in a lab; no explicit evidence
- Getting COVID-19 is a death sentence; *that's not true*

According to CDC based on a Chinese report of 18-Mar-2020, about 81% of people who are infected with the coronavirus have mild cases of COVID-19. About 13.8% report severe illness, meaning they have shortness of breath or require supplemental oxygen, and about 4.7% is critical, meaning they face respiratory failure, multi-organ failure or septic shock. The data thus far suggests that only around 2.3% of people infected with COVID-19 die from the virus.

*So, it is surely not a death sentence*

- Pets can spread the new coronavirus; *probably not to humans*
- Kids can't catch the coronavirus; *definitely, they can*

Children can definitely catch COVID-19, though initial reports suggested fewer cases in children compared with adults. A Chinese study from Hubei province released in Feb-2020 found that of more than 44,000 cases, about only 2.2% involved children under age 19.

- If you have coronavirus, 'you'll know'; *no, not always*
- The coronavirus is less deadly than the flu; *so far, it appears the coronavirus is more deadly than the flu*
- Vitamin C supplements will stop you from catching COVID-19; *no evidence*
- It's not safe to receive a package from China

For a virus to remain viable, it needs a combination of specific environmental conditions such as temperature, lack of UV exposure, and humidity - a combination you won't get in shipping packages, according to Dr Amesh A. Adalja [46], who spoke with Live Science's sister site Tom's Hardware.

- You can get the coronavirus if you eat at Chinese restaurants in other countries; *no, you can't*

By that logic, you'd also have to avoid Italian, Korean, Japanese, and Iranian restaurants, given that those countries have also been facing an outbreak. The new coronavirus doesn't just affect people of Chinese descent.

---

## 10. Lockdown a corollary to social distancing

All we described before was a process towards a long-term solution for Corona. All of that should remain glued to our minds as we explore a short-term solution, viz-a-viz lockdown.

The Indian Prime Minister Narendra Modi, displayed a great sense of awareness, while we were busy finding a needle in the haystack in the transmissions of the BBC, CNN or Al Jazeera, to make a head or tail of what exactly was happening. Why was there a commotion in the western world?

Though we were anxious about what the Indian Prime Minister might announce that day, probably something unpalatable for the citizens (like the demonetization a few years back), what he really delivered ended up as one of the greatest speeches that I have heard from him. I have heard him during elections, where all his speeches were directed to the States or the locations that he was addressing, here it was for the whole country.

He used two words that carried the essence of his talk, *Sankalp* and *Saiyyam*, He repeated these words several times during his speech, and with hindsight, I believe what he said was the crux of the issue abridged in these two golden words. He was stemming from his vast knowledge of yoga and meditation.

He said, 'this is the time for all the 130 Crores Indians to show *determination* to disallow COVID-19 entering in our lives, and *control* all our actions within the *Lockdown* faithfully.' A word that we never heard before.

We were coerced by him to be aware that irrespective of the cause or spread mechanisms of the evil, we should protect ourselves and our family members.

He closed his speech by quoting the *Lakshman Rekha*, [47] a border line drawn by Lakshman by lime on the floor in front of the entry door of the hut. Just as Seeta was forbidden to walk out of that line, we all should restrict ourselves to the confines of our homes. Corona he said, will not enter your house if you don't go out to invite her.

This speech impressed most of the Indians and its effect was seen at midnight on 22 March 2020. Barring a few urgent vehicles or people, the absence recorded on the roads was above 90 per cent.

## Lockdown Commits

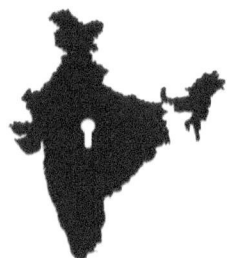

I am your saviour
Following you in the footsteps you take
Comforts in your life I make
Tiptoeing behind you
Driving the witch away

From meetings please stay away
Keep friends and foes at bay
Time will come for the hay
What do you say, dear, what do you say?
Watch them all distantly
Hear them sing remorsefully
They all sing a song of silence
They will wait if they are true
'cause they too suffer like you

I am not a physical latch
And by no means, I mean to match
That onerous corona bitch
But until you find a cure or kill
I 'll protect you and believe me, I will
Don't look for her under a glass

---

[47] From Indian Magnum opus, **The Ramayana**

Or in another biomass
She is here, undoubtedly here
The bitch is witching on a time pass
                    Outside is only politics
          Politics here politics there, politics everywhere
                    In religion and regions,
                    In the poor and the rich

I am waiting for the exit

          So, you can walk in the park with souls that vibrate
                    .cause we must all celebrate
               Barefooted on the dew drops that rest
               Later have a drink with those you trust

Trust me as I trust you

                         ******

## Lockdown as we understand

*Lockdown* is not a statement alone, and we realized it quite early. It rendered all Offices, Financial Institutions, Factories, Transports Services (airlines, trains and buses) Shopping Malls, Eateries, Pubs, Restaurants and above all the Education establishments closed.

All activities in our lives came to a standstill. And hanging over it all was Fear of Death – a creeping, invisible smog that covered more, and more, and more, of the world. We are not yet dead but the environment around us made us feel so, psychologically.

The wise will understand that the prime aim of lockdown is social distancing and isolating each other. It is a convenient and befitting method and serves as the only vaccine against the virus. It works, it worked and it helps.

Social distancing entails cancelling big planned gatherings (conferences, classes, church services, concerts, and sporting events), restricting mass transit and travel, and most importantly, working from home.

Let the truth be told, working from home is wonderful in that you don't have to adult all that much. You don't need to get dressed. And you don't need to make your bed if only you're going to sit on it all day. You needn't wait for Metros or taxis.

During this phase, the absurd, surreal limbo we can be exposed to, the physical isolation and a global digital tsunami of information that surrounds us, the loss of all the little niceties we're used to in the office or missing of a cup of hot coffee, makes us uneasy. Even the small talk in the elevator of our condominiums, and greeting each other is missing, only smiles can be seen if the mask is not worn, *sometimes*.

All this makes complaining about our lonely confinement in our homes seem like a comic cartoon as against the threat that we could be exposed to. Don't we realize once and for all, that petrol and diesel cannot be our bones of contention with the government? Transports are not our bread earners.

The Indian Prime Minister is making all efforts to stay in touch with the general public regularly to keep them in good spirits. Looking at him frequently on television screen reminds me of the days when Mikhail Gorbachev was omnipresent, almost 24 hours, on Russian television promoting *perestroika* (change) and *glasnost* (openness).

## Multiple Lockdowns – people and economy

Alongside with *Corona,* yet another pandemic emerged. It can be called an *economic pandemic.* It is caused because of *LockDown.* And we need to tackle both of them simultaneously. Not at all a mean task for any government or its people.

Good that *lockdown* is not yet rendered or interpreted as a *famine.* However, it is rather strange that the entire community that was getting fed and paid from somewhere during the *Shaheen Bagh* [48] protests from 14-Dec-2019 to 24-Mar-2020 is looking for food benefits. Where tens of thousands of protestors blocked roads affecting more than 100,000 vehicles a day, adding hours to some journeys, are queuing up to *hoard and store* rations from the Government.

The irony is seeing the flouting of the lockdown restrictions by many NGOs (Non-Government Organisations), political leaders and some benefactors inviting people at fixed clock times, to claim their rations or the cooked food, they prefer rations because they can hoard it.

Recently, it was shown on Aaj Tak television that plenty of cooked food and vegetables were stored inside homes and warehouses and were rotting, as at some other places people were dying of hunger. The food delivery systems from NGOs and Gurudwara Langar must ensure that cooked food is given on some sort of *on-line* ticket, that becomes invalid when the ticket is used once.

## How to fall in line with lockdown

### Management

The management of the lockdown project is being undertaken by the federal government and the state governments through various Cabinet Ministers and Chief Ministers, and so far, the arrangement is

---

[48] It is the southernmost colony of the **Okhla** (Jamia Nagar) area, situated along the banks of the Yamuna. It is a Muslim-dominated area.

working. But as time lapses, we will need to designate more people by name at middle and lower levels to manage things and keep it going.

Every top-level minister or executive will need committees and sub-committees to work with so the things don't fall between the cracks.

Time is ripe to test the matrix concept of management where everyone carries out two jobs, one his normal and the other his acceptance of the project virus job.

**Medical recommendations**

Clean workspaces frequently with disinfectant including high-risk areas, those that you touch frequently throughout the day.

According to WebMD, and Journal of Hospital Infection on 6-Feb-2020 [49] here is how long the novel coronavirus can live on various surface types in your home. How long it survives depends on the material the surface is made from.

- *Metal* - doorknobs, jewellery, silverware - 5 days
- *Wood* - furniture, decking - 4 days
- *Plastics* - packaging like milk containers and detergent bottles, subway and bus seats, backpacks, elevator buttons - 2 to 3 days
- *Stainless steel* - refrigerators, pots and pans, sinks, some water bottles - 2 to 3 days
- *Glass* - drinking glasses, measuring cups, mirrors, windows - Up to 5 days
- *Ceramics* - dishes, pottery, mugs - 5 days
- *Cardboard* - shipping boxes - 24 hours
- *Aluminium* - soda cans, tin foil, water bottles - 2 to 8 hours
- *Food* - coronavirus doesn't seem to spread through exposure to food. Still, it's a good idea to wash fruits and vegetables

---

[49] WebMD's Chief Medical Officer, **John Whyte**, speaks with U.S. Surgeon General Jerome Adams

under running water before you eat them. Scrub them with a brush or your hands to remove any germs that might be on their surface.

If you have a weakened immune system, you might want to buy frozen or canned produce.

- *Water* - coronavirus hasn't been found in drinking water. If it does get into the water supply, your local water treatment plant filters and disinfects the water, which should kill any germs.

One study tested the shoe soles of medical staff in a Chinese hospital ICU and found that half of them were positive for nucleic acids from the virus. But it's not clear whether these pieces cause infection. The hospital's general ward, which had people with milder cases, was less contaminated than with the ICU.

Please note the virus can live on your kitchen surfaces for three days, your countertops for over a day, and your cell phone for up to *four days*.

*How to clean and disinfect high-touch surfaces*

While many use the terms *cleaning* and *disinfecting* interchangeably, there are differences between the two.

- Cleaning refers to the *removal* of germs, dirt, and impurities from surfaces. Cleaning does not kill germs, but by removing them, it lowers their numbers and the risk of spreading infection.
- Disinfecting refers to using chemicals to *kill* germs on surfaces. This process does not necessarily clean dirty surfaces or remove germs, but by killing germs on a surface after cleaning, it can further lower the risk of spreading infection.

You should first *clean* a surface with soap and then use *disinfecting* products. If you don't have any EPA-registered disinfectant, you can use diluted household bleach solutions or alcohol solutions with at least 70% alcohol.

*For an affected person at home*

- If possible, the ill person should have one's own dedicated bathroom and bedroom. Cleaning should be limited to as-and-when required basis, to *limit* unnecessary contact with the ill person.
- Wear disposable gloves and throw them immediately after you are finished cleaning the infected person's area.
- If surfaces are dirty, clean them using a detergent or soap and water before disinfection.
- When handling laundry from an infected person, wear gloves and dispose them off immediately after touching the dirty laundry. Do not shake the dirty laundry for any reason.

## Handling the lockdown violations

While all efforts are made to tackle COVID-19 through different routes or methods, a very exceptional and most disturbing trend was seen in various parts of India. We are not aware that such incidents occur in other countries. Maybe these are not reported, so why only here?

### Troubling the medical staff and police

The doctors and medical staff were going out of the way, staking their own lives to attend to COVID-19 patients. This was really much beyond the call of their duties as they worked much more than their normal working hours. They were picking the option of working in temporary setups. They never cared for the facilities available but their concerted efforts did show astonishing results.

And now we rightly call them *medical warriors*.

The hooligans destroyed the temporary facilities and created a fear psychosis among the medical staff.

**Police in two minds**

A similar situation, but more precarious one, was experienced against the police. Police have some specific job to do, firstly to enforce that people stayed indoors, secondly to prevent crowds appearing at some alarming places, thirdly to chase in and capture patients who run away from the medical camps or the police vans, and above all to neutralize the stone pelting that takes place at many places.

India would have emerged a clear winner had these mobsters not violated the rules of lockdown. Now it's a moral question, should the police take them to task or just ensure they remain away from the infection. Till when the police would show restraint. Hooligans use flimsy grounds and justify their misbehaviour.

**The Chaos**

Certain people, groups of people, or communities from the public encircled and encompassed (*gherao*) forcibly, such noble warriors. They created disorder, administered riots and unleashed a reign of terror that couldn't be explained by anyone, even themselves.

Maybe they belonged to the opposition or different religions, but that was uncalled for.

And our government just watched and perhaps avoided proximity to the sufferers, but they could utilise the time-tested tools like lathi charge or tear gas shells. In the end, the miscreants themselves became catalysts to the increasing number of positives of COVID-19. What a shame and what inaction!

In particular, a couple of cases deserve to be listed here.

**Violations – promoting the virus spread?**

The social media mentions at a few places that Christians abhor the lockdown when the rules are applied to churches or their surroundings. But the way Muslims have opposed it in India pulls out the rug from under my feet.

### Tablighi Jamaat – the religion

Tablighi Jamaat is an Islamic missionary movement that urges Muslims to return to practising their religion as it was practised during the lifetime of the prophet Mohammad, and particularly in matters of ritual, dress and personal behaviour.

Followers must dress like the Prophet, sleep as he did on the ground, on one's right side; enter bathrooms leading with the left foot, but put pants on leading with the right foot; do not use a fork when eating instead use your index finger, middle finger and thumb; men shave their upper lips but let their beards grow; their pants or robes should be above the ankle 'because the prophet said letting clothes drag on the ground is a sign of arrogance'.

The Tablighis move in cities, towns, villages, stay in the mosques and engage with the Muslims of the locality. They tell the young minds to distance themselves from the world, education, career and indulge instead in the opium of religion. Inference of this is: there is no need to open your shop, earn money, plough field. Allah would take care of everything, even of the children they give birth to.

They inform the participants about what happens after death, in the *grave*, and the *afterlife* and how to prepare for those eventualities. They propagate the worldly life as too insignificant to be cared about. Once a Tablighi, nothing interests him except religions and its precepts. Mostly, lower-class Muslims, who see no hope of making good on the mainstream, seek achievement and redemption through this route.

### The origin

Established in 1927 by Muhammad Ilyas al-Kandhlawi in the *Mewat region of India*. He became a missionary for reforming Muslims but didn't advocate preaching to *non*-Muslims. Later he relocated to Nizamuddin near Delhi, where this movement was formally launched.

Tablighi Jamaat denies any affiliation in politics and jurisprudence, and focuses on the Quran and Hadith, as long as they don't deviate from the Sunni Islam. However, the group adopts a *non-violent* ideological stance. Tablighi Jamaat defines its objective about the concept of *Dawah* and interprets *it* as *enjoining good and forbidding evil* only

So far so good.

## Its spread

The organisation is estimated to have between 12 million to 80 million adherents worldwide in about 180 countries, with the majority living in South Asia. It has been deemed as one of the most influential religious movements in 20th century Islam.

Its first foreign missions were sent to the Hejaz (western Saudi Arabia) and Britain in 1946. In France, it was introduced in the 1960s. The United States followed and during the 1970s and 1980s, the Tablighi Jamaat also established a large presence in continental Europe.

It might interest some of our esteemed readers that Pakistani professional cricketers Shahid Afridi, Mohammad Yohanan, Saqlain Mushtaq, Inzamam-ul-Haq, Mushtaq Ahmed, Saeed Anwar and Saeed Ahmed are active Tablighis.

## Literates behaving as Illiterates

The Tablighi arrogance comes from the belief that they are doing God's work and God is their refuge. Care and caution, the advice of the medical practitioner, a directive of the government are not in their dictionary since all that goes against their concept of how life is to be led.

Tablighis believe coronavirus is a curse of God let loose on the sinful humans to teach them a lesson; and a solution to this problem is indulging in religious rites, prayers, and visiting mosques. The answer to this virus threat is not closing the gates of mosque/church, but *opening* them to allow people to bow before the altar of God.

## Clash of religious beliefs

Though quite a many Muslims associate themselves in India with the activities of the Tablighi, which is a puritanical movement, they oppose it tooth and nail. They call it *Taklifi Jamaat*. They see it as an enemy of the Muslims, since it persuades the Muslims to renounce the world, and take to prayers and propagate the puritanical form of religion as a full-time activity.

## The chronicle of events during Lockdown in 2019–20

*South-East Asia*

In Feb-Mar-2020, the Jamaat organised an international mass religious gathering at a mosque in Sri Penta ling, Kuala Lumpur, that resulted in more than 620 COVID-19 cases, making it the largest-known epicentre of the virus in Southeast Asia, at that time. Brunei, Singapore, Thailand, Cambodia, Vietnam and the Philippines came into its envelope.

*Pakistan*

Tablighi men organised a congregation of 250,000 participants near Lahore on 11-Mar, despite the authorities requesting them to postpone or cancel it. But the participants had already gathered and communed together. So, they gifted the perils of coronavirus and deaths to the whole of the Muslim world, from Gaza to Malaysia.

Raiwind, the place where the event was held was locked down by Pakistani authorities. Later, as luck would have it, a Tablighi stabbed a policeman while trying to escape from the isolation facility. He was later arrested in Khyber Pakthunwa while the policeman was hospitalized in Layyah town.

*India*

The Tablighi Jamaat bid for holding the jamaat was called off by the State Government of Maharashtra. This timely response saved the residents from a devastating disaster.

The Nizamuddin faction of the Tablighi Jamaat, however, went ahead to hold its congregation (Ijtema as it is called) in Nizamuddin West, Delhi. It was held in every week of March till 21-Mar-20.

The Delhi Government's order of 13-Mar, that directed against the holding of seminars, conferences or any big event (beyond 200 people) was ignored by the organisation, and the Delhi Police also failed to enforce it.

167 Jamaatis were quarantined after evacuation from the Markaz (centre), in a railway facility in south-east Delhi. The Tablighi Jamaat gathering emerged as one of India's major coronavirus hotspots undeclared at that time. Finally, 1445 out of 4067 cases were linked to attendees according to the Health Ministry of India.

On 31-Mar-2020, a First Information Report (FIR) was filed against *Muhammad Saad Kandhlawi*, the grandson of Muhammad Ilyas al-Kandhlawi who is heading one faction of this organisation.

**The probable reasons why Tablighis misbehaved**

The behaviour of Tablighis is a question of belief, faith and religious doctrines. If we ignore their belief, we could easily become the herd of the ongoing confusion against the sect, as the television has. If we as literates give consideration, we could be on the neutral side.

Certainly, there are rogues in every sect, cult or religion including the Hindus, Christians and Muslims but haven't we tolerated them despite their crimes like rapes and murders. Let the truth be told, we want them all to be locked up in jails and wait for the inevitable, the prolonged confinement that might ultimately result in death.

Of course, spitting on fruit, licking the saleable fruits, passing the kissed currency during transactions, pelting stones on the Medicare attendants and police who came to their rescue, sometime in the presence of a magistrate, throwing urine-filled bottles and breaking them, misbehaving with nurses, dancing naked are actions which no person would approve, whether or not the COVID-19. And this was carried out throughout the country.

How can Allah be pleased by these unsocial and violent activities for which the country's law will ultimately hold? It is much against what Allah wants. He just wants to switch off the businesses and return to a puritan's life. But violence is no way. They cannot justify their actions from any reasonable pedestal.

In the background of what CDC and WHO have stated, there is a possibility of one individual infecting up to 59,000 people if we consider 10 cycles of cascaded transmission. If Jamaatis are spreading this disease to south-east countries, they should also choose who lives and who dies. But they are simply and universally striking at the root causes of their own economies, including hotels, bars, restaurant and massage parlours.

But a question has been planted in my mind, 'why is Mr Saad not arrested so far', in spite knowing very well that he cannot change clothes because of Allah's instructions, and his house being no less than a palace?

And now I believe he is sending messages on video clips from an unknown place. Are our intelligence network and all its satellites so weak that he cannot be detected and located, or is the will still lacking form the government? Are we thinking of future elections?

## Mayhem at Ahmedabad, Surat, Bandra Mumbai etc

On the day when Mr Modi extended the lockdown until 3-May, ie on 14-Apr-2020, a sudden crowd assembled at many cities' railway stations. These were presumably the labour class in amazingly large numbers, mostly from UP wanting to go back to their homes. The news about the extension of a lockdown or otherwise was spread as an irresponsible rumour.  But what we cannot understand is why it was not refuted by the government

- only the Muslim workers wanted to go back to their homes in UP
- why they assembled near Jama Masjid near Bandra
- why no woman was visible in the thousands present
- why there was no baggage in their hands
- why the location selected for this congregation was 3 kilometres away from the railway departure terminal
- above all why the presence of an organised loudspeaker in the hands of the announcer
- and why at the same time in different cities spread over 500 kilometres away?

## Commentary

For me, the answer lies in inadequate communication and implementation. Because the void of information is quickly filled by speculations and rumours, mainly through social media. The people are eager to hear not only what the prime minister has to say, but what their chief ministers are planning. For example, If alcohol is an income source, then let the states decide on it.

Bring back the workers from their villages ....

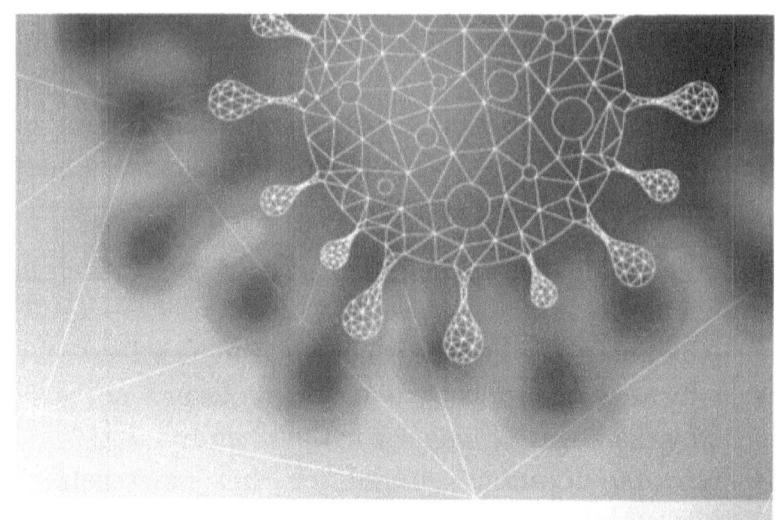

# THE
# CORONA
# SYNDROME

A chance for introspection

MIKE RANA

# 11.    The economic write-off

Lockdown was sudden and nothing in it could have been preplanned. Quoting the Indian Prime Minister, as he said 'if 21 days lockdown is violated our country could go back by 21 years'.

There is a hidden threat under this statement. The clouds of economic battering hover over our heads, not knowing what to do or what the government can do. The economists are tearing their hair out at the thought of what lies ahead. And this ahead is not a few months spill over but could extend to years.

This is a completely different approach than the US-based, *Anthony Stephen Fauci,* who proposes a loosened approach, but defines a logical method.

### Unemployment and released labour in the USA

The fallout of the corona pandemic took a brutal turn after a record of 6.6 million Americans filed for unemployment recently. Those eye-opening numbers trump the previous record – 3.27 million jobless claims – which was filed just the week prior, according to the Department of Labour.

Nonetheless, the fact is that America had led its public in the habit of *borrowing* and *spending* with the start of the credit card system, a bubble that took America, the largest economy of the world to a whole new level of *vulnerability*. The example Daniele Booth quotes in an interview is that a credit card invitation came to her *6 months* old son in the email.

Such a trend indicates that Americans may be afraid to take care of their day-to-day affairs if their cards don't work.

A very interesting discussion on the eroding values of Federal Reserve vis a viz the American psyche of buying without the support of their own money appears on YouTube. (refer to the link given in the footnote) [50]

A few trillion dollars may now be consumed in clearing (or carrying forward) their debts. The story in Russia and China is the same. So, the economic blunder blues in lockdown may not be worse than what you and I think.

**Devastation in India**

No such study was made or released by India, but it couldn't be better. Worse if we count in the street beggars. [51]

As soon as the lockdown was declared, exodus occurred. People had no option but to head back to their hometowns or villages. Hoping they would at least be close to their near ones, and a possibility of food grains becoming available or could be grown. Hopefully, the impact of the virus spread could be less within the fewer travelling people in villages. A somewhat compromise of a solution.

The students were asked to vacate hostels as facilities ran short and couldn't be maintained. Elsewhere, it broke your heart seeing the bartenders, ushers, UBER drivers, flight attendants, yoga instructors, barbers, and hotel cleaning staff being pushed under. Wipeout two, three, or six months of income from the booking clerks at your local theatre, or the griller at your local bar, and you're *impacting generations.*

But a bigger problem existed. No transports (cars, buses, trains, or airlines) worth a mention were seen or arranged. On the way, they had all the risk of being intercepted by the police fulfilling their

---

[50] https://www.youtube.com/watch?v=jsV_YXq-1x4 the YouTube interview

[51] As per the minister's statement, a total number of **4,13,670 beggars** are residing in India, including 2,21,673 males and 1,91,997 female beggars in the country. The data is based on 2011 census

duties. The media was full of viral videos of what ill happened because that's what always attracts human attention.

The dissatisfaction, hopelessness, helplessness and anger writ large on their faces, reflected how they had been, so-to-say, cheated by their employers or the government. They wouldn't have left the places where they were working just for nothing.

Just imagine what might be going on in their minds when they started walking on foot to cover a distance of hundreds of miles to reach their villages, with all their scanty belongings on their head packs in bales, impacting adversely the psyche of children. With no food arrangements. And in thousands. They were forgetting that ultimately, they themselves could become the carriers of the virus for their own so-protected village!

Still, the *biggest* dilemma as realized by all of us now, that we never thought of earlier, is how on earth can social distancing be implemented if 8 to 10 people live together in one room of 8 by 10 size, in many places in India. A blatant case of Dharavi suburb in Mumbai widens our eyes.

And where are the homes for people living on the roads?

And at some locations, the police were seen giving small punishments to drive home the point that they should remain stuck to their homes.

Lockdown – Could have been like in Vietnam one leg locked in

The Indian Prime Minister was forever trying to coax and educate the masses to understand that lockdown was the *only* option in front of us, as long as the vaccine was not on the horizon, despite all our inconveniences.

**Some alarming figures**

The following small list enumerates some difficulties of the people and their employers;

- Look at the public at Bombay Victoria Terminus, for that matter any other railway station or bus stands in any city. The sight of people sitting on roofs is quite common. And most of the area close to terminus or bus stations are forever overcrowded.

  Sometimes I wonder if people are moving for work or going back to villages, to celebrate festivals, to enjoy their holidays, or because of reduced (almost nil) job opportunities.

- Hotels and tourism industry employ 4 Crore [52] people. Their forecast of job losses is 12 Lakhs [53] and Rs **11,000 Crore** revenue loss
- Aviation industry revenue is 2.2 Lakh Crores [54], 3.5 Lakhs people, and they expect a revenue loss of **4,200 Crores** in two months alone
- Retail industry of 59 Lakh Crores, expects a 4.6 Crore and job loss **1.1 Crore** in just 3 months
- Restaurant industry 73 Lakh people, expect a job loss of **14 lakh**
- Real Estate expects a job loss of **35 %**
- OLA and Ober services that provide earnings to about 5 Million drivers, 50 to 60 % loss is already registered but will increase manifolds with unitised services
- Major car production units were shut down and labour went back to villages
- Most offices were shut down and workers retrenched
- Squeezing of non-organized sectors (sanitary workers)

**The Black Monday on 20 Feb 2020**

While the world stock market was moving towards correction for the last few months, a global financial crash was seen from 24 to 28-Feb. Stock markets the world over reported their largest one-week declines since the 2008 financial crisis.

The early March was set aside by extreme volatility, with large swings occurring even in day-trading. On 9-Mar-2020 most global markets reported severe contractions, mainly in response to COVID-

---

[52] One Crore is 10000000 with 7 Zeroes = 10 Million

[53] One Lakh is 100000 with 5 Zeroes = 0.1 Million

[54] One Lakh Crore is 100000 0000000 with 12 Zeroes = 100 Billion

19 and an ongoing oil price war between Russia and the OPEC countries.

Oil prices had dropped in 12 weeks by 68 %, compared to 100$ in the preceding 72 weeks. You might as well know that two-third of oil goes into transportation and the COVID-19 lockdown reduced this demand to just 10 % all over the world.

A stock market crash is a social phenomenon where external economic events combine with crowd philosophy in a positive feedback loop where selling by some market participants drives more participants to sell.

Acts such as wars, large-corporation hacks, changes in federal laws and regulations, and *natural disasters of highly economically* productive areas may also influence a significant decline in the stock market. The stock markets currently being down is also because of a never-ending lockdown as seen by the investors.

But remember all such stock drops may result in the rise of stock prices for corporations competing against the affected corporations.

Now we should be worried about what will happen to the oil market when the lockdown is released. The owners of oil countries will play their cards shrewdly, Russia and Saudi Arabia may become our biggest culprits. The US has started preserving its oil barrels.

**The Government declares freebies**

Freebies distribution is a compromise resolution that the Indian government emulated from other countries. It's the only answer in the current hardships. Public money has vanished in thin air. Some of it may have been transferred or deposited abroad by unscrupulous people or is tucked away in the pockets of privileged people in India.

Haven't we seen photographs of richness on the covers of Forbes and Time magazines? Haven't we seen how the rich celebrate weddings or anniversaries? The money is all here and it can be easily seen and marshalled if the government has the will to do it.

The Indian government is short also because of the forecasted collections. Leaving aside our doubts about the veracity of the published figures on Google, the following may interest you.

- From the expected collection of GST of 7.4 Lakh Crores, the government received 5.8 Lakh Crores which is short by over a **Lakh Crores**
- From the expected collection of Income Tax of 5.2 Lakh Crores, the government received 4.7 Lakh Crores which is short by **50000 Crores**
- The expectation on the GDP growth is at a very low at 4 % even in the next year 3 % will not be registered. But that was well before the lockdown started.
- Indian stock market is currently down by 35 % in two months

Interestingly some foreign countries like USA, Canada, France and even tattered Italy have dished out substantial packages to tackle public demands in the lockdown period, and a majority of it is in the form of *direct* transfer to people's accounts.

In the first lot, the Indian Government dished out a package Rs 1.7 Lakh Crores. For the time being let's leave out the possibility whether it will reach the desired people and what happens to those who aren't given a package but are very much on the radar of the lockdown, for example, the small industrialists and traders.

Without going into the specific details, scanning the Finance Ministry's freebies report by a microscope at the global level, we may be able to locate the meagre relief for people. It is understandable keeping the huge population in mind.

Naturally, the benefits would go to the below the Poverty Line People (BPL), whereas the part of the relief may propagate to the middle class as well, in the form of *deferred* payments, though.

People who pay Income Tax and General Service Tax (GST) have received a token relief of extended dates for tax payment by a few months, deferred linking of PAN cards with Aadhar Cards, deferring EMI payments for about 6 months for home or car loans or payments to the Insurance scheme.

The irony is that most of the schemes are just fiddling with people's own money, like loans, EMI or Employee Provident Fund (EPF). All benefits will have to be returned after the show is over.

And latest we hear is the holding back the raise on dearness allowance for the central government employees which was announced earlier. Another date will be declared later. Further for many relief clauses, the government has roped in third parties like State Governments and Reserve Bank of India. For the delays and deferred reliefs, 'the ball is in your courts, friend'.

**Government adopts an unkind posture**

Let's talk about the traders and small-scale manufacturers. The above picture indicates their plight.

Government charges 12 to 28 % GST as soon as a trader raises a bill, and it is purported to deposited with the government; irrespective of whether the bill is paid by the recipient. Further, on his investment, the trader pays 18% GST for loans taken by him.

So far so good, it's the expected norm and has been pushed down their throats for a fairly long time. But, as the lockdown eases out, the government expects them to pay the salaries of the absence of labour, and also bear the fixed charges for electricity when the factory was not even run - which might be reasonable because that is what expected of everyone.

But what kills the traders and manufacturers is the obligation of the owner to bear the increments of revamped daily wages and cost of food for the deployed labour. The argument is that the labour is expected to stay within the premises due to lockdown.

It is in complete contrast with what is being done for medical warriors or the sanitary workforce. Only a few are accommodated inside the hospitals, not all.

Wait, the hammer is still in the air. When it falls ie when any of the above-mentioned lapses is noted, an FIR may be stamped against the owner.

Isn't it far-fetched? Have a heart for the traders, they too have to survive in this situation. They are working round the clock to make good the losses they incurred during the lockdown. It's their time to turn the screw and prevent their balance sheet not going into the red.

How long will the traders or manufacturers be able to sustain their businesses? Is this a plan to support the Chinese takeover of the industry? If only the government agrees to the following, the trading community will glow to brilliance.

- Pay the salaries to the labour for the period of lockdown

- Waive off the electricity fixed charges since Mar-2020 until the factory reopens
- Should reconsider the mandatory requirement of boarding and lodging of the employee/labour within the premises

## Reinforcing the food banks to keep feeding Millions

### The schools' program

The Mid-day Meal Scheme is a program of the Indian Government for improving the nutritional standing of school-age children.

It supplies free lunches on working days for kids in primary and upper primary classes in schools falling under different categories approved by the government. It serves 120 Million children in over 12.65 Million schools and Education Guarantee Scheme centres, and they say it is the largest of its kind in the world.

Distribution of food is a common issue that inherits some idiosyncrasies in each location. Since COVID-19 is far from over, national and local groups are focusing to do their best in two directions. To ensure that children don't remain hungry when the schools are locked out. And to guarantee that children in schools don't go without food.

Food was an issue in the minds of people, not only in parents but the public at large. The Supreme Court Chief Justice of India SA Bobde took initiative and issued notices to all State Governments and Union Territories, asking how the government will provide mid-day meals.

Some state governments provided a temporary answer to the Supreme Court. Maybe because they were engrossed much more in providing support to the millions of daily wages workers, and their families. But generally, all states responded quite well, to the best of their capability.

Some of the states distributed money to the accounts of people, some arranged packed food to be delivered and some others arranged raw food grains to each parent of children. At places, people arranged to collect from central designated places, and at some locations, food or grains were delivered at the doorstep.

**Common solutions related to food grains and cooked food**

This was and is going to be an unsurmountable issue in the future, and needs to be tackled differently and innovatively.

- **First**, the management of Food Corporation of India (FCI) warehouses must be re-tuned to ensure that food grains don't continue to lie on floor and rot, (rat and rot menaces) during the humid weathers in the country.
- **Second**, the distribution link in the chain of stock approval of the government-approved fair price ration shops should be eradicated from its prevalent rampant corruption. A lot transpires between the trader, local MLA, to the Central government minister who allocates the stock. Much can be said about this problem but the paucity of space inhibits its description.
- **Third**, the arrangements should be streamlined for the food banks or food courts to have varying price tags, package types, and locations. All these must be encouraged and regulated with complete control on the way food is cooked, stored or delivered. This is a very professional type of activity and cannot be ignored.
- **Fourth**, the distribution methodology of rations and cooked food for Below Poverty Line people or for the public under distress needs to be revamped. This should be recorded

specifically and state-wise, in the Disaster Management manuals of the government.

**Food Banks**

With COVID-19 shuttering businesses across the country, the need to feed the hungry grew rapidly. And we can't ignore the regular lunching requirements of the office goers, which is gaining grounds off late.

The way food banks will work will depend on how agreeably the food bank chains will operate with the grocery stores or the farmer partners. All of them need to work out details on pricing, and marginal pricing (ie price when the demands vary drastically with seasons or trends).

Farms have excess produce, and they need and would welcome additional delivery chains. That's the most exciting part of the whole process. So far it has been ignored. It's a global business opportunity, but mind you it may need setting up of MEGA food factories.

Then there is the question of tying up with larger food chains such as Zomato, for which the newcomers will have to upgrade to the product specifications and the consistency desired.

Such kinds of associations or agreements make sense to get monetary donations instead of food, in disasters. With the same money that people contribute, five or six times more produce can be purchased by the foodbanks.

Not bad really.

**What we learnt from Corona**

- Chine won the third world war without firing a single bullet
- Westerners are fewer followers of rules except when we penalise them, Indians are a submissive lot
- The rich are less immune than the poor

- No astrologers, priests, tarot readers or gurus saved even one individual, not even on a mental or emotional plane
- Health professionals are worth much more than a football or cricket players
- Oil is a worthless commodity when there is no consumption
- Animals in zoos feel exactly how we feel during a lockdown
- The planet regenerates more quickly than human efforts, except we must let it
- Majority of people can work from home
- Everyone can survive without junk food
- Men can cook too
- Media is full of non-sense, nothing they say is reliable
- Actors are just entertainers and not our heroes

## Hope sustains life, literally

Being housebound, hunched over an office chair, stuck in tiny rectangles of zoom isn't what typical well-being is meant to be. I'm feeling it, you're feeling it, your colleagues, peers, clients, reports, bosses, and competitors are feeling it. But we are all waiting for the virus to go away automatically. It won't happen unless some bright chap finds a vaccine. But now there is hope.

With no new cases reported in Goa, Tripura, and Manipur and one patient in Arunachal Pradesh getting cured of coronavirus, these four states have become coronavirus-free even as the total number of coronavirus cases in the country approached 27,500. This reminds me of the famous song of 1964 by The Animals - The House of The Rising Sun, why because these **states** represent the states of the rising sun in India.

Bombay and Vadodara were the cities that never slept, these days they turned into cities that never wake up. The streets are bare in Chandni Chowk, or Crawford Market and no one is making prayers at Gurdwaras, Temples or Mosques. All cultural monuments are closed. If you're at the moment in Bombay, the 9/11 at Taj seems very little. But there is hope.

Until then, I suppose all we can do for now is to fight back in the small battles of our daily lives and turn inconveniences into positives. For example, I am plying Accordion much more time though causing irritation to my daughter and wife, who would like the Netflix (now even at SD format). Perhaps we should work out more, eat better, spend time with our families, and take a rest from our busy, busy lives (when we're not waiting in queues at Spenser's, Reliance, Big Bazaars or tying outfits at Marks & Spencer's).

the Animals - The House Of The Rising Sun 1964 (High Quality)

## 12.   Unwinding the lockdown - the exit

The first question that strikes us is whether or not we will unwind the lockdown. If yes, when. We can all have our guesstimates but no one, even the Prime Minister, can answer that question. It will depend on where we stand on our economic front or our dying list.

Dr Ashok Kapur of Chandigarh who is handling patients for over 40 years in his hospital, made a rather alarming assessment. He said, 'Coronavirus is not going away, it will keep coming back year after year like flu, according to a new study.'

My assessment is that its return will be stronger and all-encompassing depending on how strong our immune system is or how strong we can make it through the use of vaccines. So better take care of your immune system, by whatever means you know of. As normal citizens, without any special powers endorsed upon us, our concerted opinion is we'll never return to the original life settings, at least for a few years.

However, only 25 % of Indian population is keen on getting back to the old-world order, and because of their comparative older age. They like to take us back to the past. But others, the 75% are youth, who are uncertain and insecure will not let this happen. Because they will hold the steering wheel in their ruthless hands.

After the unlock, the situation might radically transform, with businesses putting up false façades of doing well. A small business owner may lose his bread and butter. Another entrepreneur may replace him. A lot of uncertainty exists in the real estate, ghost towns are glaring at us glass-eyed, waiting for the angels and vampires to dwell in them. And we know the youth is not planning to own properties or cars. They are once bitten twice shy. So the real estate will stay put unless the government does something worthwhile.

And not to forget, our tour operators, the airlines, railways and the cruises will not bloom up abruptly. They need cranking up.

Although the government has a big problem of managing the drooping GDP, they will have to awaken the sleeping giants on the Indian soil. 'Make in India' slogan must re-emerge recklessly. The start-ups have a golden opportunity up their sleeves.

One call is rampant in the horizon, not to let the Chinese take over our economy and the big giants.

**Don't throw away the positives, please don't Mr Modi**

**Freedom of animals** - It was in the news, and you must have also heard it. A leading politician mentioned about Rs 1500 to 2000 Crores loss due to fall in exports of non-vegetarian (maybe Cow's meat) in Feb- 2020. In Economic Times (7-Mar- 2020) it was mentioned that only 70 / 240 Month on Month containers were exported in this period.

Nothing can please me more than these statements. Not because I don't care about the economy but only because we inadvertently have given animals a chance; a chance to make their presence felt, a chance to share the world stage with us, they have an equal right to breathe unobstructed air and above all escape the tortuous untimely death.

Ducks were seen on the landing strip in Israel, Nilgai (Blue Bull) on roads in Noida, Dolphins, who like the freshwater, were seen at Marine Drive Bombay. Even in China, that deals with animals rather harshly, in some cities eating cats and dogs has been banned.

A picture of Blue Bull, Neel Gai clicked by me behind our home in Gurugram.

**Reduced pollution levels** - Low pollution levels due to the missing air conditioners and traffic were conspicuous in all cities. The skies were clear – blue rather than grey, and as someone said you could see the statue of liberty from the top of the hotel La Meridian in Delhi. Jokes apart, reports were pouring in from the city of Jallandhar in Punjab that the far stretched (250 KM) white snow on the Himalayas was out in the sun and inviting us over.

And in most of the northern hemisphere in India children could experience how the stars, *twinkle* like a litter star.

The emission of Carbon Dioxide and Carbon Monoxide and Nitrogen Dioxide was recorded at lower numbers. Cut down of construction activity, both buildings and roads contributed to a cleaner dust-free environment, but carrying it too far is going against progress and is neither possible nor desired. Instead, we should learn from China to mechanise a lot in the real estate industry.

**Waters** - Surface water, including streams, rivers, lakes, reservoirs, and wetlands have shown an astonishing improvement in their composition. All the money that our governments spent on Ganges cleaning, was not desired if we were to change and implement stricter laws on effluent disposal. As for the affluent, it is thick and turbid and stinks like a dirty brewery. [55]

Did you Mr Modi see the Ganges water surfing and cleaning automatically just because chemicals were not poured in it from Kanpur. This is the real gain of lockdown; a tighter control on the industries.

We can't say the same for groundwaters (22 % of the total waters) but extrapolation proves that these too must be emanating cleaner and returning to their original characteristics. Hope in future we can drink directly from the taps.

**Crime reporting** - Less crime rate in particular rape, and traffic violations were recorded. It must be a welcome change. We must now work in our new world towards unsocial elements not being tolerated and our jurisprudence guiding our judges justly.

But how will we control domestic violence and children abuse that must have escalated during the lockdown? The answer is in tightening the law but keeping options for both sides to be heard.

This is also the time when we must keep a very tight vigil on hackers who have found a god-sent opportunity to engage in *online frauds*. This might increase a lot more, as the government pushes the pedal on coronavirus-related freebies.

**Urinating and spitting on walls and roads** –Drastic reduction is seen in these wrongdoings but that's because people are locked indoors, not because their mis-adventures were curbed. I shudder to think that this big improvement will vanish as the lockdown unwinds.

---

[55] Certain things must be removed from the **wastewater**. This includes organic matter, inorganics (sodium, potassium, calcium, magnesium, copper, lead, nickel, and zinc), pathogens, and nutrients (most notably nitrogen and phosphorus). The treated wastewater can then be safely discharged into water bodies, applied to land, or even reused in plant operations.

This is something that we need to learn from the British. Whereas the British penalised people who spat on the Mall road Shimla, and the signboards for that effect are still there, the Indian government doesn't even object to advertisements for products like Gutka, Chutaki and Pan Bahar. Disgustingly, a very prominent ad for Pan Bahar remained glue to the television screen as The Indian Prime Minister, a founder of Swatchh Bharat mission, addressed the nation on 14-Apr-2020.

I happened to see the wholesale market in Jaipur Rajasthan recently, and maybe there are other places too, where these disgusting products are sold. And what more, its mafia earnings go into making of movies in Mumbai, which are frequented by our youth.

Golf Course Road in Gurugram, Haryana is a prestigious and posh road with 6 channels on each side and 5 underpasses that provide *traffic light less* drive for over 5 kilometres within the city. There is a blank stretch about one kilometre long that probably will be covered by a bridge over a creek. (The photograph of Neelgai was clicked by me here). When? God knows.

It is a convenient spot for peeing for the drivers who happen to be driving along that stretch. Believe me, one cannot pass on this road even on the opposite side as the drying urine pools stink on this stretch. But things aren't so good on the other side as well. The modern plaza and the footpath stretches are always painted red by spitting of the Gutka.

The list goes on and on, for the goodies that we experienced. While the country is on the mission of Swachh Bharat, we should force public urinals to be built on long stretches of very many roads.

**While we unlock the lockdown, Mr Modi please carry on and insist on one-week junta curfew every month.**

## Everything need not be declared wide open

They can open up the facilities to some degree and in phases, but the risk of a rebound will continue to haunt us until we get a very broad-spectrum vaccination.

Bill gates had put in 3 Trillion dollars on the anvil. But according to this great philanthropist, 'If all the necessary interventions are imposed things won't return to normal until the Fall of 2021'. However, very recently mention has been made that like the H1N1 virus the coronavirus can't be eliminated.

Most of the state governments in India seem to be liking the concept of lockdown, they are in favour of continuing and they wouldn't want anything to moderate the goings-on. Yet, prudence says something must be done to keep the labour on their bread and the economy on its feet.

We can't let down the untiring efforts of doctors and nurses; what if the virus re-appears.

The book attempts to make some suggestions, which the government may or may not follow.

### Professions-wise grouping for easing out

All economies work with different professionals and everyone from these groups needs a different type of working arrangement, some critical, some important and some pure profit-making. Here is a representative list

- Medical doctors, nurses, paramedic and helpers
- Police, paramilitary and even military
- Agricultural segment including their workforces and farmers
- Manufacturing industries, Heavy, Medium, Light industries generally run by individuals
- Transport industry all modes, state-owned and business owned

- Service and distribution industry, on-line or others
- Banks, post offices and insurance companies
- Education industry teachers, students and supports thereof, focus preferably on on-line education
- Food industry recently became popular
- Information Technology, the ultimate enablers

**Product and service-wise grouping that is vital**

- Online home delivery of services like Amazon, Flipkart etc including antiseptics, foods and napkins. Stricter controls are required for packaging materials and social distancing at the time of direct delivery
- Home delivery of cooked foods from designated sources and liquor (to ensure excise inflow to the government).

Only such food delivery services that offer a *contactless delivery* option should be permitted. The delivery courier will leave the food at the door to be picked by you after he leaves. This way, social distancing can be maintained.

The bags, boxes, and foil used for packing the food items are not essential. These should be thrown away immediately. Treat it like dry waste and take it out to the nearest garbage bin where it can be picked up.

To prevent contamination of your food, wipe the plastic containers with sanitizer or disinfectant before opening the lids. Transfer food into washed and sanitized home vessels, and dispose of the plastic containers.

Don't forget to wash your hands after bringing the food into your house, and again after throwing away the packaging and containers.

- Medicines
- Transportation like trucks, seaports and associated labour

The idea is not to be perfect in this classification. But it should be understood that no compromise is done on the Social Distancing aspect.

**Sectors of the population as Epicenters for Corona**

The biggest confusion that may surface will be the changing nomenclature of the status for containment. The central government, state governments and the media might confuse people, by using variable names. For example, Epicenters, Hotspots, Red, Orange, Green labels topped up by Containment may be the terms used for referring to them. How can the 1.33 billion people follow any guidelines if these status nomenclature keeps on changing.

And India is a large country the authorities must use a common framework for naming the sectors. The best and easiest seems to be Red, Orange, and Green labels. The span of these labels may be confined/restricted to physical locations such as pockets, areas, locations, sector, district or taluka that could be controlled or locked down.

A *hotspot* (red, orange, or containment) is the one that shows the gravity of infections, about 10 people affected as *positives*, during the last fortnight from the date of monitoring or examination. It's a moving average, meaning the situation may change daily.

Once declared a hotspot, it will remain so for a long time before returning to normal. Remember, even one individual can cause the virus spread, and the system will get *rewound* to the situation of lock-down.

**Setting priorities**

The priority should remain the health of people, the economic considerations in that order. According to this paradigm, all those who are involved in health care are in any case cleared to continue for work.

Out of the rest, daily earners will be allowed to start their activities. Second on the list are home delivery persons, postal services, banks, sweepers employed in vital lotions, security guards, drivers, service or maintenance engineers etc. The guiding principle should be to reduce the suffering of daily earners so that they can earn their livelihood.

If at all some relief is granted to a taluka or sector, for which separate specific announcement will be made, it will remain in vogue only and until all its members follow the guidelines or rules for the Lockdown. If these conditions cease to exist, the lockdown will be reintroduced for that taluka, street or sector.

Meaning it is the responsibility of people squarely to retain their status as a green label and show their literate attitude.

More ...

- Open liquor shops 2 *consecutive* days a week
- Some dedicated effort is required to repatriate the Indian living abroad from their current locations, but this requires to be done with stricter pre-flight tests and should be done *flight by flight only*. And when they arrive in India, they will be quarantined for 14 days.
- Similarly, some arrangements for students returning to their home states must be planned by the state governments. For example, all buses earmarked for such transfers should be completed sanitized before allowing students to enter. And after they arrive at their destinations, the students hopefully will propagate the rules of social distancing in their home towns, or villages.
- Allow domestic airlines with a full physical check of passengers with a dip test or litmus kind of early detection. Introduce an escrow account receipt for a deposit of Rs 10,000 per trip to reduce the travels. Take a date validity

certificate from every traveller that he or she is void of any coronavirus.

- Permit opening of 4 to 5 stars hotels with reduced strength since travels will remain restricted and much better discipline can be expected in these hotels cum resorts
- Open the courts at 50 % level services. Those that can conduct court in-camera or by video conferencing should be preferred to function
- Permit real estate service but make sure that only on-going project can be undertaken and no new projects. Focus on sale first

## Country's economy

Irrespective of what has been said, we can't be fools tuning down our economies to absolute zero even though there might be dangers to some lives. We need an ongoing system, that can be recuperated easily and sustained when the virus is gone. In this respect, I completely agree with *Anthony Stephen Fauci MD*.

'If you knock down the economy completely and disrupt infrastructure, you may be causing issues and their unintended consequences, with a *risk of no return*. Elderly, please stay out of society in self-isolation. Don't go to work if you don't have to. No bars, no restaurants, no nothing. Only essential services.'

## The specifics for a partial opening-up

With the rider of Social Distancing

- No unauthorised person (and all of us are unauthorised) shall move or travel outside the house without wearing face masks or id / Aadhaar cards (if outside your resident's society)
- Only one or two persons per house will be allowed outside the house at a time for a specific purpose and for not more than three hours at a time.

- No gathering for any purpose more than five persons, no religious congregations and all places of worship shall remain closed.
- All those who have been permitted to work should have a *green spot on the screens of their telephones* that states the corona status of the individual, and which can be shown on demand.
- Work should continue in offices with 25% attendance with masks and gloves
- No person above the age of 65 and with any history of comorbidity (hypertension, diabetes) or undergoing any treatment for cancer or major ailments should be permitted to move outside the house. Such individuals should be given special passes if needed
- An over-riding condition for private vehicles will be to restrict their movement as per the well known odd-even scheme. And no private vehicle movement will be allowed during *Sunday*.
- Attendance at marriages and funerals will be restricted to 20 persons. But a prior approval may be required from the District Magistrate
- State borders remain closed other than for emergencies and essential products
- Supermarkets, malls, theatres, bars etc with closed air-conditioning should *not* be permitted to be reopened.
- In this phase, inter-district bus transport with protocol restrictions may be allowed and so will the domestic flights for essential passengers, doctors, health workers and patients.
- *Online* sale of liquor may be started since this brings in sufficiently large excise returns for the government, much needed at this time.

## The middle class too expects some favours

In the light of precarious situation for the government, people have shared a lot of losses by sitting down at home, but a few things should fall into their laps as well

- Some expenses need to be cut off. Haven't we agreed to pay the laid-off labour, our maids, drivers their salaries
- All commercial electricity bills or fixed charges be cut to half for the period of lockdown stating 1-Apr-2020
- Companies should be allowed to retain 50 % of GST payable for the next 12 months
- Interest payments should be waived off for the next 6 months, just like the government does it for the farmers
- All EMI's to banks and NBFC to be put on hold for 6 months with no levy of interest on delayed payment
- The employee share of the PF / ESIC not to be paid by the companies but to be borne by the government for 6 months
- Property tax for FY2020-21 to be reduced to half for all properties

**The return to normal**

It is our misery that we never believed in collecting data for subsequent analysis so that we could see the trends, particularly symptom wise deaths or inflictions.

We are now living in the times of what is known as Big Data. And we have a specialised profession for this activity. I fail to understand why is the country not collecting or publishing the data for analysis?

Why do we have to extrapolate data collected by USA or elsewhere?

A positive interview of HDFC Bank MD Mr Aditya Puri was held recently about why India will survive and make it big

- India's rural economy is not affected by corona and it's going on strongly
- India being a young people's nation won't have much detrimental health effect of corona compared to Europe
- Merchants and small shops are not over-leveraged (not too many loans) so will emerge strong once shops reopen

- India's internal consumption is strong, after all, we are 1.3 Billion people
- Even if the stock market collapses and has to enforce few circuits closures its mainly because of automated algorithms which are forcing stock selling and nothing to worry
- Work-From-Home (WFH) is reducing costs for all companies and may increase profits eventually.
- Recently the Information Technology company the Tata Consultancy Services (TCS) announced that as much as 75% of its strength will work from home until 2025.

**A unique paradigm of lockdown – herd immunity**

The world is going one way in dealing with coronavirus, Sweden the other. No lockdown, only restrictions of gatherings of more than 50. Schools are open, public transport is running, cafes and bars are open with the restriction of table service only rather than crowding at the bar. But don't forget it's for the citizens to follow the rules meticulously.

An advisory, and just that, has been issued for those above 70 to maintain social distancing. And people have only been advised to try to work from home as far as possible; no shutdown across workplaces.

Sweden is not coronavirus-free. Far from it. It has reported more than 18,640 confirmed cases with 2194 deaths. Sweden has a population of just above 10 million.

The Swedish government has taken this path as considered policy, on the advice of government scientists though more than 2,000 of them have signed a petition asking the government to bring in tighter measures over social distancing.

Chairman of the Nobel Foundation Prof Carl-Henrik Heldin issued a statement about what the government was doing, last week of Mar-

2020, saying 'we have let the virus loose... they are leading us to catastrophe.'

This is the path that Britain had set out on, before a group of 229 scientists warned the government that such a policy could lead to up to half a million deaths by Aug-2020. The British government swung into an *abrupt U-turn* following that warning. And all of Britain now lives in the fear that its government could have lost precious time, and reversed its ways rather too late.

The Swedish government has built its policy on the argument also that no lockdown can be maintained long enough and full enough to stop the spread of the virus physically. Blanket prevention is not possible, only a managed spread is. That reasoning now faces the test of time in Sweden and the rest of the world, in opposing ways.

In respect of confined or restricted clusters, we could follow a *herd immunity* paradigm. Under such a policy the virus should be *allowed* to spread. The logic is that this will not affect most people seriously; those who get it, and get over it, can then get on with their lives. And others remain aloof out of the herd.

**An encouraging note on reviving our economy after Corona**

Don't take me for granted but if our government works with prudence, and be stricter on lockdown violations

- Our Lockdown, hot weather and our higher immunity levels will see us through.
- We can become the largest exporters of medicines. The world will look to us for regular supplies of BCG vaccines, anti-malarial drugs Hydroxy Chloroquine etc.
- Make-in-India, manufacturing facilities will be set up in India, in preference to China, by corporations from all over the world.
- India can become the hub for the manufacture of every item from mobile phones to pharmaceuticals

- Against the threat from the animal kingdom, our vegetarian cuisine will be accepted
- Our medical facilities will be appreciated for their easy availability, quickness and price efficiency.
- The corona may help to reverse the brain drain

**Commentary**

'Go to the polls,' 'go back to work' and 'go buy my anything' all sound more and more like 'it's safe, go outside,' as summer approaches.

Fifty-nine per cent of the Gen Zers polled in and said another four weeks is all it will take for them to lose their minds, and 80% said that they have the emotional capacity for 2 months of lockdown at best.

'While there are massive societal costs from the current shelter-in-place restrictions, I worry that reopening certain places too quickly will almost guarantee future outbreaks and worsen longer-term health and economic outcomes,' Facebook CEO, Mark Zuckerberg said in a recent media release.

No, I think we'll have to wait until it's all over. Right now, it's almost like the fog of war. After the war is over, you then look back and say, 'Wow, this plan, as great as it was, didn't quite work once they started throwing hand grenades at us.' It is similar to that.

The coronavirus is certainly scary, but despite the constant reporting on total cases and a climbing death toll, the reality is that the vast majority of people who come down survived it.

US and India are positioned on opposite sides of the equation economy versus lifesaving. Even Sweden has adopted the business stand, and the UK has turned turtle from business to life-saving. Who is right and who wrong the time will tell? And it will be interesting to know who makes the reverse turn first.

The WHO has stated lockdown and the business will have to go hand in hand if countries have to survive. Which seems to be right.

So, keep your businesses intact, look for new avenues like WFH and Video Conferences, certainly real estate and roads will not be done on these two media. But a great change is expected in the education services with more and more text and examinations and degree available online.

*Jaan hai to Jahaan hai*

is the call of the time

and not 'Shutdown the Shutdown'

**A comment from a friend in Russia**

COVID-19 you have been heartless,
I can easily call you a serial killer on the loose
In your murderous spree
You have managed to unite people and nations
To create borderless allies
You've managed to bring down narcissistic powerful leaders
Who thought money can buy you a way out of death?
You've managed to give planet earth and wild animals a break, You gave a fucking wake-up slap across the face to EVERY living human across the globe.
That's a major accomplishment you fucking virus 🦠

And one thing is for sure,
We will overcome this and live to tell a great story
With major Hollywood blockbuster movies

**A comment from Zacharoula Christou from Cyprus**

In the year of COVID-19

I will always remember this time and think of this virus as a great, son of a bitch, bastard, heartless teacher

But still will remain in my heart as a teacher despite all the gruesome adjectives which come along with it.

It strengthened my belief that we ONLY have this fucking life!!!!

It made me even more adamant that we need to be taking care of ourselves first to be able to take care of others

It made me appreciate that we need to be adventurous and be able to accept that NOT everything goes to plan and that's fucking OK

It made me value my God-given right of freedom and appreciate more how animals or even other humans must be feeling in captivity

It really made me realize the value of a handshake, of a strong hug and of an emotional kiss

One thing is for sure, I will not be waiting for the weekend, or a special day or some vacation or holiday break to enjoy living anymore and I will instil this notion upon my children

I will be smiling even wider, I will be hugging more, I will be experiencing more with my inner core, along with people who share the same beliefs as mine

www.ingramcontent.com/pod-product-compliance
Lightning Source LLC
Chambersburg PA
CBHW021415210526
45463CB00001B/377